ROUTLEDGE LIBRARY EDITIONS:
COMPARATIVE URBANIZATION

Volume 4

LAND AND HOUSING POLICIES IN EUROPE AND THE USA

LAND AND HOUSING POLICIES IN EUROPE AND THE USA

A Comparative Analysis

Edited by
GRAHAM HALLETT

Routledge
Taylor & Francis Group

LONDON AND NEW YORK

First published in 1988 by Routledge

This edition first published in 2021
by Routledge
2 Park Square, Milton Park, Abingdon, Oxon OX14 4RN

and by Routledge
52 Vanderbilt Avenue, New York, NY 10017

Routledge is an imprint of the Taylor & Francis Group, an informa business

British Library Cataloguing in Publication Data
A catalogue record for this book is available from the British Library

ISBN: 978-0-367-75717-5 (Set)
ISBN: 978-1-00-317423-3 (Set) (ebk)
ISBN: 978-0-367-77201-7 (Volume 4) (hbk)
ISBN: 978-0-367-77206-2 (Volume 4) (pbk)
ISBN: 978-1-00-317023-5 (Volume 4) (ebk)

Publisher's Note
The publisher has gone to great lengths to ensure the quality of this reprint but points out that some imperfections in the original copies may be apparent.

Disclaimer
The publisher has made every effort to trace copyright holders and would welcome correspondence from those they have been unable to trace.

LAND AND HOUSING POLICIES IN EUROPE AND THE USA:
A Comparative Analysis

EDITED BY GRAHAM HALLETT

R

ROUTLEDGE
London and New York

First published in 1988 by
Routledge
a division of Routledge, Chapman and Hall
11 New Fetter Lane, London EC4P 4EE

Published in the USA by
Routledge
a division of Routledge, Chapman and Hall, Inc.
29 West 35th Street, New York, NY 10001

Printed and bound in Great Britain by
Biddles Ltd, Guildford and King's Lynn

British Library Cataloguing in Publication Data

Land and housing policies in Europe and the USA:
    a comparative analysis.
    1. Land use, Urban 2. Urban policy
    I. Hallett, Graham
    333.77'17   HD111
    ISBN 0-415-00511-6

Library of Congress Cataloging-in-Publication Data

ISBN 0-415-00511-6

CONTENTS

LIST OF TABLES AND FIGURES

Tables

Figures

BIOGRAPHICAL DETAILS

Graham Hallett is on the staff of the Centre for European Community Studies, University of Wales, Cardiff, Great Britain. His books include The Social Economy of West Germany, Housing and Land Policies in Germany and Britain and Urban Land Economics: Principles and Policy.

Richard H. Williams is Senior Lecturer in Town Planning, University of Newcastle-upon-Tyne, England. He has published articles on a range of European planning themes, and is the editor of Planning in Europe (1984).

Barrie Needham is an economist working in the Faculty of Geography and Planning at the University of Nijmegen in the Netherlands. He has published widely in English and Dutch on land policy and its relationship to town planning.

Jon Pearsall is Senior Lecturer in Government and Planning, Chelmer Institute of Higher Education, Chelmsford, England. He contributed the chapter on France in Housing in Europe (ed. M. Wyne, 1984).

Georgia Grzan-Butina, who was educated in Yugoslavia, the USA and Britain, is a Lecturer in Environmental Design at Oxford Polytechnic, Oxford, England.

David E. Dowall is Professor of the Department of City and Regional Planning at the University of California, Berkley. He has published extensively on urban development in the USA and has recently completed a land market study of Bangkok for the National Housing Authority of Bangkok.

ACKNOWLEDGEMENTS

A grant under the Working Parties and Specialist Conferences Scheme of the Nuffield Foundation made possible several meetings of the contributors to this book. Without this financial assistance, the project would have been less successful and might not have come to fruition at all.

Chapter One

INTRODUCTION

Graham Hallett

This book arose out of a feeling that countries could learn (either what to do or what not to do) from the urban land policies of other countries, but that there were few international comparative studies available - and that those which covered more than two countries were rarely satisfactory. This is perhaps not surprising, given the inherent difficulty of producing an international comparative study on any subject.

One method is for an individual to write about a number of countries. A single authorship makes possible a unity of style and approach, but few people can be really knowledgeable about more than two or three countries. More common is the 'symposium' by national experts, usually the proceedings of a conference. This often suffers from the lack of a common approach. A person writing about his own country tends implicitly to assume an understanding of institutions and concepts which may be completely alien to outsiders. The consequent confusion is frequently compounded by problems of language. Probably the best arrangement is for a group of people to work closely together, with one of them acting as editor. I have been fortunate to find a number of specialists on the land and housing policies of various countries, who agreed to participate in a joint (and unfinanced) effort. We hope that we have avoided some of the pitfalls of the 'symposium'.

Subject Matter. One question which exercised us was how much ground to cover. We wished to concentrate on urban land policy - rather than 'housing policy' - because this wider context of housing policy has been neglected by mainstream academics. A new study also seemed appropriate because the 1980s have, in many countries, seen traumatic changes in economic conditions and associated land use. The 1980s are an era - like those which gave rise to the doctrines of Henry George and Karl Marx - in which the rich are getting richer and the poor poorer, and in which changes in

1

the ownership of, and access to, real estate have contributed to this 'polarisation'. The results are evident in homelessness and decaying 'inner cities' alongside million-pound homes.

Land policy, however, is closely connected not only with housing policy, but with town planning, taxation, local government organisation and other issues. If we had confined ourselves to land policy in a very narrow sense, the result could have been somewhat arid, since land policy is not an end in itself, but a means to ends such as well-planned cities, a minimum standard of housing for all, and an equitable distribution of income. On the other hand, there was a danger of spreading our net too wide. We have therefore tried to concentrate on 'land policy', but to say enough about housing and town planning to clarify the effects of land policy on city development and the housing situation. Different contributors have given different slants to their chapters, reflecting the particular problems of their countries, but we have tried to cover a common 'core' of topics, especially:

(a) the buying, selling and holding of land by public agencies.
(b) the land market, including the impact of taxation and subsidisation.
(c) the control of the land market through town planning controls, etc.

In this list, the term 'land' has been used in the extended sense in which it is used in 'land law' i.e. real estate, covering both virgin land and land with buildings on it, when the question of redevelopment arises (with particular reference to housing).

Value-free Science? Land policy has historically been a controversial subject, and still is. According to the 'positive', or 'value free' school of social science, however, the social 'scientist' should be as objective as the mathematician, taking the line, "My studies show that if the policy-makers want 'a' they should adopt policy 'x'; if they want 'b' they should adopt policy 'y'. I have no views as a social scientist, whatever views I may have as a citizen". An alternative view, put forward in a neglected book on the methodology of economics (Walker, 1943), is that value judgements inevitably influence writers' approaches to a subject, even if they believe that they are being completely objective. It is therefore better – while trying to be as factual and objective as possible – to be open about one's economic and political philosophy, rather than pretending to Olympian detachment, and to examine 'normative' alongside 'positive' issues.

My personal observation of the eye-gouging feuds of 'value-free social scientists' has strengthened my belief that Walker's view is correct. I must therefore admit that I am, in the words of the Chairman of the Planning Committee of the (late) Greater London Council, 'another bloody liberal'. The same could probably be said of the other contributors. We accept that the New Right, and the New Left, have put forward some ideas which are of value in the reassessment of urban policies. The New Right is correct to point out that political intervention is undertaken not by philosopher-kings but by officials and politicians, whose interests may not coincide with those of the public, and who may be 'captured' by those they seek to control (Littlechild, 1986). Moreover, the price mechanism and private ownership have great virtues. But the argument that a market economy can solve all urban problems, and that state intervention is necessarily harmful, is not one which our study of urban history leads us to accept. We find equally unconvincing the New Left's arguments that these problems would vanish with the public ownership of all land and housing; this does not seem to us to be supported by the experience of Stalinist countries. We conclude that a mixed economy in land and housing is both demonstrably possible, and desirable. There is a role for private ownership and the market, and a role for public action; the ultimate purpose of studying different national systems is to define the two roles.

An Apologia
My original interest in international comparisons was stimulated by the time I spent in West Germany and Canada, where housing and land policies were less of a 'political football' than in Britain. When, in the mid-1970s, I wrote Housing and Land Policies in West Germany and Britain and Urban Land Economics (which were denounced, or ignored, in British academia because of their 'Right-wing bias'), I believed that Britain had taken some wrong turnings in its policies on 'council housing', rent control, 'comprehensive redevelopment' and 'the nationalisation of development rights', and could usefully learn from West Germany and Canada, and where the virtues of a market economy were combined with decentralised state intervention in the housing and land market.

There followed the 'Thatcher revolution'. It (soon) seemed clear to me that the new 'radical Conservative' policies were as doctrinaire as any of the earlier Labour policies - and at least as harmful. Any reader of the British chapter who disagrees with the authors' critical assessment of the Thatcher Government's record in the field of land policy, housing and town planning must discount their 'Left-wing bias'.

Table 1.1: Basic National Statistics

	Population 1984 (million)	Area (000 sq.km)	GNP per capita 1986 ($)	Average inflation rate 1973–84 (%)	Projected population growth 1980–2000 (% p.a.)	Urban population as % of total 1984 (%)	Average growth of urban popn. 1973–84 (%)
USA	237.0	9363	17,240	7.4	0.7	74	1.3
W. Germany	61.2	249	14,540	4.1	-0.1	86	0.3
France	54.9	547	12,740	10.7	0.5	81	1.2
Netherlands	14.4	41	11,700	5.9	0.4	76	-1.0
UK	56.4	245	9,600	13.8	0.1	92	0.2
Yugoslavia	23.0	256	n.a.	24.6	0.6	46	2.7

Source: World Bank. World Development Report 1986

Housing Tenure*

	Owner occupied	Rented		Average house** price 1984	No. of hours work needed to buy house** 1984
		Privately	Housing Assoc. public body		
	%	%	%	£	
USA	65	32	3	53,000	6,400
W. Germany	40	45	15	84,000	20,453
France	51	26	23	29,820	9,928
Netherlands	44	13	43	32,660	8,339
UK	62	8	30	31,160	9,165

* 1982-84 Estimated   ** Nationwide B.S. 'House Prices in Europe' (1987)

INTRODUCTION

The Countries

International comparisons of one aspect of public policy are most illuminating when the countries compared do not differ too much in other ways. West Germany, the Netherlands, and France are all 'Roman Law' (and Carolingian) countries, with many similarities. These countries are grouped together, starting with Germany, which is my starting point in international comparisons. The next chapter is on Yugoslavia. Yugoslavia might seem an 'odd man out' but I have been struck by the extent to which the Yugoslav experience illustrates several recurrent themes. We then come home to Great Britain, before turning to the USA which, as David Dowall explains, cannot be regarded in the same light as any single European country.

A few basic statistics of the countries concerned are given in Table 1.1. Yugoslavia has an exceptionally low urban population as a percentage of total population, and has had an exceptionally high growth of the urban population in recent years. Yugoslavia is thus experiencing the type of problems which the other countries experienced in the 1950s and 1960s. The forecasts of population growth from 1980 to the end of the century shows that there will be a slight fall in West Germany and a slight rise in the UK; in the other countries, population will continue to grow, although at lower rates than in recent years. The demographic trends in Germany and the UK will tend to ease pressures in the housing and land markets, although they also create a demand for different types of housing.

There are some notable differences in housing tenure between the countries. In terms of owner-occupancy, the USA and the UK are at the top, with percentages in the 60s, the Netherlands and West Germany are in the 40s, and France occupies an intermediate position. Yugoslavia does not have national statistics, but the figure in the cities is between 30 and 60 per cent. 'Public housing' - using the term in a broad sense to include subsidised housing provided by the state or housing associations - ranges from a mere 3 per cent in the USA to 43 per cent in the Netherlands. The figures of 'hours of work needed to buy a house' should be interpreted with caution. They suggest, however, that the 'real' price of owner-occupied housing is lower in the USA, and higher in Germany, than in the UK, France and the Netherlands. However, homeownership is not (yet) the norm in Germany, and the average quality of houses is higher than in the other countries.

Cause and Effect

Any attempt to discover, by means of international comparison, the effects of specific policies encounters a familiar difficulty. Social mores, legal systems, and 'national character' all play a role in what happens, so that it is difficult to isolate the effect of public policies in general and, still more, the effect of any one policy. Only tentative conclusions can be drawn, and perhaps the main benefit lies in the process of reaching them, rather than in the conclusions themselves. If people study the range of policies and problems in other countries - and other times - they can escape the growing tyranny of current fashion, blinkered perspectives, and 'what I say three times is true'; they will be in a better position to decide what ought to be done, even if they still have differing views.

The national chapters are solely the responsibility of the respective authors, and I take full responsibility for the final chapter.

REFERENCES

Littlechild, S.C. (1986) The Fallacy of the Mixed Economy, Institute of Economic Affairs, London

Walker, E.R. (1943) From Economic Theory to Policy, Chicago

Chapter Two

CRITERIA FOR THE ASSESSMENT OF URBAN LAND POLICIES

Graham Hallett and Richard Williams

Any assessment of urban land policies faces two problems. The first is that of isolating the effect of a range of public policies - which may sometimes have contradictory effects. The second is that of deciding - implicitly or explicitly - the criteria for deciding the success or failure of past policies, the nature of current problems, and the aims of future policy. It is sometimes possible to examine these issues in relation to the stated objectives of the governments concerned, and so avoid using any 'subjective' criteria. But official objectives are often couched in such vague and unexceptionable terms as to be meaningless. A review of a 'Greater London Development Plan' gives a delightful list of its stated objectives, beginning with - 'to give new inspiration to the onward development of London's genius' (HMSO, 1973).

Even when they are not meaningless, however, the objectives endorsed in public by politicians are rarely the whole story. Indeed, there are theorists at both ends of the political spectrum who argue that all statements of public policy objectives are bogus. Marxists argue that (under 'capitalism') politicians and civil servants - and mainstream academics like ourselves - merely reflect the interests of the bourgeoisie. The 'economics of politics' theorists, on the other hand (who tend to support laisser-faire), argue - or at times seem to argue - that everyone is purely self-seeking (Buchanan, 1978). One may concede that professed ideals are often tinged by vested interest. By reviving the ancient Roman question cui bono? ('What is in it for them?'), the 'economics of politics' school has performed a useful function, but its more rigorous exponents arrive at conclusions which are contrary to everyone's knowledge of human nature. We believe (with Adam Smith) that self-interest is an enduring motivation, which should be utilised for the public good through the 'invisible hand', but that people are not purely selfish. People do have ideals, and ideals are, in the long run, influential. It therefore seems worthwhile trying to

8

compile a list of objectives for urban land policy, which are widely acceptable - and certainly acceptable to the authors. Those with different objectives can discount our 'bias', and use different criteria for assessing the national policies described in the following chapters, but clarifying one's objectives is useful in itself. Our discussion will be taken up again in the concluding chapter.

British and German Objectives

A convenient starting-point might be a comparison of the objectives given in two 'Building Land Reports' by the West German 'Ministry of the Environment' (Bundesminister, 1983, 1986) with those in the British (Labour) Government's 1965 White Paper, The Land Commission (HMSO, 1965).

The German reports listed the objectives of municipal land policy as;

1. facilitating the implementation of land-use plans.
2. providing cheap land for income groups which would otherwise be 'squeezed out'.
3. preserving open space.
4. bringing unused land into use, and so contributing to a generally careful use of land.

The British Government gave the objectives of what became the Land Commission Act as;

'1. to secure that the right land is available at the right time for the implementation of national, regional and local plans;
2. to secure that a substantial part of the development value created by the community returns to the community and that the burden of the cost of land for essential purposes is reduced' (para.7)

The first objectives in both lists are much the same, but 'plan implementation' requires some elaboration. Land can become available through market mechanisms as well as through compulsory purchase. A 'shortage' of land can, however, arise - even when enough is zoned for development - in two ways. Firstly, physical, administrative or economic obstacles may prevent land (especially land which has already been built on) being developed; derelict factories or mineral workings can remain for years, or decades. Secondly, land may become available only through an 'excessive' rise in price (which yields 'supernormal profits'). A surge in demand for land, impinging on 'inelastic' supply, can thus produce a 'price explosion', such as occurred in 1973/4 in several countries. Property booms of this type are short-lived, but they have undesirable consequences, not least that the sight of 'speculators' making large gains can lead to the adoption of ill-considered policies. We therefore accept the first object- ive. In so far as it is achieved, land prices will remain reasonably stable, and land will be made available to meet demand (within the planned framework) as a result of market reactions, public intervention, or some combination of the two.

Betterment and Compensation

The second paragraph in the 1965 White Paper raises two related issues; 'betterment' and the cost of 'land for essential purposes'. The reference to betterment tends to confuse two distinct cases because of an ambiguous use of the term 'community'. In some cases, land prices rise as a direct result of expenditure by public authorities, e.g. a new road. It seems reasonable that the authority concerned should, if it is feasible, be able to recoup some of this rise in land values. In other cases, development value is not 'created by the community', at least in this sense; it is simply the result of economic changes. It nevertheless seems reasonable that a proportion of development value should be taxed, subject to the 'canons of taxation' - that the tax should not have un- desirable side-effects, should not be unduly expensive to administer etc. Some development value is necessary if land is to be sold voluntarily; the question is how much can be creamed off without seriously affecting incentive. The ques- tion also arises whether a special tax is necessary; it may be preferable to rely on general tax provisions, or on an annual property tax.

This 'public finance' approach (very different from that of the fundamentalist 'land reformers') underlies two reports on the taxation of development value in Britain, which contain suggested criteria for a satisfactory tax system (British Property Federation, 1976; RICS, 1974). These criteria include;

'1. There must be an equitable division of development gains between the Treasury (i.e. the central government), the local authority and the landowner/ developer.
2. The tax system should not discourage landowners from making land available for development.
3. The administration of the system should not over- burden government, and should not be dis- proportionately expensive.
4. The system should be consistent with a mixed economy and permit an adequate rate of investment in development, especially from institutional sources.

5. It should be such as to command broad political support.'

We accept these criteria, which are far from being truisms, since they have all been explicitly rejected at various times.

The Gains of Owner-occupation

The reports quoted above discuss only 'development gains'. Although they may be the most striking, these are not the most important gains from the ownership of real property. Owner-occupied housing provides an 'income in kind' and, with owner-occupancy ranging from 40 to 65 per cent in our sample of countries, is now a major factor in national income and capital stock. In 1985, the value of 'personal sector' dwellings in Britain - £240bn. - was equal to the total market valuation of British registered companies. The value of all residential buildings in 1985 has also been calculated as 36.5 per cent of national capital, up from 26.1 per cent in 1957 (HMSO, 1987).

The sale of houses (in some high-value districts) is also becoming a significant source of income. As long as a person occupies a house, or sells it and buys a similar house, the level of house prices is irrelevant. But people who 'trade down' when their family has left home, or who inherit a house which they do not need, can in some cases make a large gain (Morgan Grenfell, 1987). The tax/benefit system should, we suggest, ensure a 'reasonable' level of taxation of the income (in cash or kind) arising from homeownership - again bearing in mind feasibility, administrative cost, and equity - and should concentrate benefits on people with lower incomes (which is the reverse of what is happening in some countries).

Land for Essential Purposes

The 1965 White Paper refers, in addition to 'betterment', to 'the burden of the cost of land for essential purposes'. Its implicit assumption is that a local authority should be able to buy land cheaply, but this view can be challenged on grounds of both efficiency and equity. If, for example, local authorities can expropriate urban land for motorways at well below market value, there will be no direct incentive to minimise the loss of housing involved, and the narrow perspective of the highway engineer is likely to predominate. Similarly, if public bodies can expropriate housing with little or no compensation, in the name of 'slum clearance', there is likely to be (indeed, in Britain there has been) hardship to some owners, and an unwisely ruthless destruction of older housing.

11

The general point is that the price of real property reflects 'opportunity costs' - however imperfectly - and thus provides a guide to the use of resources. On the one hand, local authorities should not be 'held to ransom' by individual property owners, holding strategic sites which can block development; compulsory powers are needed as a last resort. On the other hand, draconian powers of expropriation at well below 'market value' can be inequitable; can encourage excessive and undesirable acquisitions of property by public authorities; and can sometimes be self-defeating because of the political opposition aroused. The procedures for compulsory purchase, and the basis of compensation, should strike a balance between these conflicting considerations, rather than treating compensation merely as a 'burden'.

Who Gains; Who Loses?

Land policy cannot be separated from wider issues of town planning. As well as ensuring that land is available for permitted development, it is also necessary (if one accepts the case for any kind of town planning) to ensure that un-authorised development does not take place. But there is an important qualification. Unauthorised development must really be harmful, and the plan must take adequate account of the demand for housing. Allowing people to obtain what they want is, in the end, more important than 'plan implementation'. In some countries (including Yugoslavia), illegal building is a major industry. Since, however, it enlists resources for housebuilding which would otherwise not be available, the question arises whether the authorities should not legalise and guide it, rather than somewhat ineffectively trying to stop it.

In other countries the problems are different, but the same basic question arises, 'Is the plan actually in the public interest?'. Since, however, 'the public interest' is notoriously difficult to define, it might be better to ask, 'Is the plan contrary to the interests (as they see it) of significant num-bers of individuals?'. It is now generally accepted that some of the plans adopted with such confidence in the 1950s and 1960s (e.g. British 'slum clearance' or the French grands ensembles) were, at least in part, misguided. In response to criticism, town planning in these countries has become more sensitive to what people actually want - as distinct from what the authorities think they should have. In the USA, on the other hand - which until the 1960s had very little public control of land use - there has been a move (now partially reversed) towards greater control. The controls were purportedly for environmental reasons, although the results were sometimes the opposite of what they were supposed to achieve.

Since town planning is a political activity, the first question to be asked of any plan should be, 'What are the intellectual ideas and interest group pressures which have contributed to its making?'. This question has not always been asked. In the post-War era, plans in some countries tended - in the words of one enthusiast-turned-critic - to be treated as 'absolute moral imperatives, almost entirely un-related to any discussion of the aspirations and needs of the mass of the people' (Clawson and Hall, 1973). It is now more generally understood that plans are both subject to architect-ural and intellectual fashions, and also the outcome of political processes, in which the interests of groups lacking political influence may be ignored.

This raises the question (No. 2 in the 'German' list) whether public authorities should make land or housing avail-able to low-income groups which would otherwise be 'squeezed out'. 'Disadvantaged' groups can suffer both from the unre-strained operation of the market, and from some planning policies, e.g. 'comprehensive redevelopment' and 'exclusionary zoning'. The fact, therefore, that there are many more gainers than losers from a particular plan is a necessary, but not - we suggest - a sufficient, condition for its adoption. Efforts should be made to identify those who are likely to lose as a result of the implementation of the plan. If the harmful impact can be removed by modification to the plan, the plan should be modified. If it cannot, then the losers should, as far as possible, be compensated with alternative land or housing, or with cash. Compensation is particularly important when the losers are poor, but a general presumption in favour of compensation can be justified on grounds of both equity and expediency. We would, however, distinguish between an actual loss and a hypothetical loss resulting from limits on an owner's right to develop his property in the most profitable way. This distinction between becoming poorer and not becoming richer is implicitly accepted in all the countries studied, which all provide no compensation for limitations on development which are part of a local plan. But there is still a strong case for the judicious use of 'sweeteners'.

The Objectives of Land Policies
To summarise our argument; urban land policies should be judged according to the extent to which they;

 (a) provide an adequate supply of land for housing,
 (b) facilitate good town planning (which is perhaps more easily recognised than defined),
 (c) improve access to land and housing for 'dis-advantaged' groups,
 (d) impose 'reasonable' taxation on the gains from the ownership of real property,
 (e) compensate property-owners (and tenants) for actual losses resulting from town planning policies.

All three principles have at times been rejected. Principle (e) was rejected in parts of British post-War legislation on land and housing. At the present time, principles (b) and (c) tend to be rejected by 'the New Right'. The view is increasingly heard that any assistance to the poor only makes their situation worse; that the best way of helping the poor is to help the rich - or at least to reduce the fiscal disincentives to becoming rich, and the 'welfare' provisions which cushion the effects of being poor. These New Right views have in recent years been backed up by assertions of the type, 'We are all Friedmanites now'. In fairness to Professor Friedman, it should be pointed out that, in a famous book, he rejected this attitude to poverty (Friedman, 1962). Although eulogising 'capitalism', he accepted that, by itself, it would not eliminate poverty, and proposed doing so by substituting a 'negative income tax' for housing subsidies and other 'welfare' measures.

As a substitute for the jungle of separate welfare provisions, a 'negative income tax' has considerable attractions, but there is still a case for 'subsidising houses' in certain circumstances. Land prices in some districts are so high as virtually to exclude lower-income families. If it is considered that such districts should not be confined to the very rich (to provide 'social mix' or to prevent the creation of 'dead' commercial districts) there is no alternative to subsidising either land or housing through a public, or non-profit, agency. Making land or housing available cheaply to owner-occupiers is not a permanent solution, since their successors will have to pay the market rate. Even if a 'negative income tax' were in existence, it would not obviate the need to consider the consequences of major land, housing, and town planning policies on the distribution of real income among different social groups. In fact, Friedman's twenty-five year old proposal has not been introduced in any country, and finds no support whatever among the Governments which profess to support his ideas. The 'radical conservative' Governments have cut back 'welfare' payments and public housing subsidies without introducing an alternative system for the relief of poverty.

We conclude that, under a 'capitalist' system, land and town planning policies should have, as one objective, that of helping the casualties of the system. There are however, others. One report lists the purposes of town planning as follows (Nuffield, 1986).

'(a) To monitor and control the impact on the environment of present and future uses of land.
(b) To anticipate and prevent the perpetration of nuisances.
(c) To provide a coherent and consistent framework for the operation of the market.

(d) To reconcile conflicting demands for land as they arise from the development plans of private and public agencies.

(e) To assist in the promotion of whatever developments public and private are considered desirable by the relevant public authority.

(f) To provide the information necessary for the effective discharge of these functions.'

This list reflects the prevailing chastened mood of British town planners; they present themselves as traffic policemen rather than as guides to Utopia. Supporters of a 'mixed economy' like ourselves can welcome the list - as far as it goes. But the authors are perhaps too anxious to insist that planning is 'value free', and (e) smacks a little of the doctrine that a public servant has no responsibility to the public.

Urban Renewal
From the 1950s to the 1970s, urban land policy in most of the countries studied was concerned with 'greenfield' development. With the decline in population growth and the emergence of 'inner city' problems, the emphasis has shifted to 'urban renewal'. This shift, combined with the end of the age of full employment, has necessitated considerable re-thinking of town planning and land policy. Our basic criteria in the assessment of 'renewal' policies are;

(a) the extent to which they impose a brake on the downward spiral of decline, and encourage private redevelopment and renovation (using 'private' to include cooperatives). Public investment is often needed to 'prime the pump' but it is unlikely to succeed in the long run unless it is followed by private investment.

(b) the extent to which they improve the lot of the poor and 'disadvantaged' local population. A 'renewal' which simply forces the poor to another area is not a satisfactory solution.

Centralisation versus Decentralisation
Finally, the theme of centralisation versus decentralisation runs through our survey of land policies. Town planning and land management are in some ways best carried out at a local level. On the other hand (as the authors of the chapters on the USA and Yugoslavia suggest), there is a case for 'broad-brush' land-use planning over a whole conurbation. A more far-reaching question - raised by 'Thatcherism' - is whether planning and land management should be delegated to elected

local authorities or controlled by the central government or its nominees. Some light may be cast on this issue by the experience of countries ranging from the highly decentralised to the highly centralised.

CONCLUSIONS

There are many possible ways of judging urban land policy, based on considerations of equity, efficiency, and political or social desirability. The subject is therefore controversial, but we have outlined the objectives which provide our criteria. In some cases, it is possible to ascertain with reasonable precision whether an objective has been attained, e.g. stability of land prices. In most cases, however, objectives cannot be used as precise measuring-rods. Terms like 'an equitable division of development gains' or 'disproportionately expensive' or 'good town planning' clearly leave room for argument. Moreover, one objective may run counter to another; it is thus a matter of weighing up several considerations to produce a reasonable balance, rather than discovering the 'ideal' policy. Any policy which is put forward in capital letters as 'The Solution' (HMSO, 1965) or because, in Mrs Thatcher's words, 'there is no alternative' should be treated with some scepticism.

REFERENCES

British Property Federation (1976) Policy for Land, London

Buchanan, J.M., et al. (1978) The Economics of Politics, Institute of Economic Affairs, London

Bundesminister fuer Raumordnung, Bauwesen und Staedtebau (1983 and 1986) Baulandbericht, Bonn

Clawson, M. and Hall, P. (1973) Planning and Urban Growth; An Anglo-American Comparison, Johns Hopkins Univ.

Friedman, M. (1962) Capitalism and Freedom, Chap. XII, Chicago

HMSO (1965) The Land Commission, Cmnd. 2771, London

HMSO (1973) Greater London Development Plan: Report of the Panel of Inquiry, Vol. 1, London

HMSO (1987) Economic Trends, May, 'National and Sector Balance Sheets, 1957-85', London

Morgan Grenfell PLC (1987) Housing Inheritance and Wealth, London

Nuffield Foundation (1986) Town and Country Planning: The Report of a Committee of Inquiry, London

Royal Institution of Chartered Surveyors (1974) The Land Problem - A Fresh Approach, London

Chapter Three

WEST GERMANY

Graham Hallett and Richard Williams

INTRODUCTION

Urban land policy in West Germany is firmly centred on local government, and has been characterised by considerable continuity in its evolution. It has been closely linked with both the post-War housebuilding programme and the extensive programme of urban renewal carried out in recent years.

Measured by results on the ground, West German land and housing policies are generally considered to have been extremely successful. In 1945, West Germany faced a housing situation incomparably worse than that of any of the other countries under review. Up to half of the housing in most cities had been destroyed, and a catastrophic shortage was aggravated by an inflow of refugees, which eventually totalled 13 million. Yet today housing conditions are among the best of the countries concerned – especially if one looks at the poorest section of the housing market. There is virtually none of the dereliction and homelessness found in the worst districts of US and British cities, and most foreign visitors to the 'problem districts' of German cities wish that they had such minimal problems.

Admittedly, this favourable verdict is not always shared in German academia. The German contributor to a recent symposium describes the post-War achievement as 'something of a myth'; talks of a 'new housing crisis' and the 'poor state of housing in West German cities', and seems to advocate 'council housing'.

> 'The validity of a policy to radically reform and re-structure the Federal German 'social' housing system in the direction of council-owned or municipalised housing can be further argued from the standpoint that the existing forms of municipalised housing in other West European countries arose out of similar 'crisis' developments' (Kennedy, 1984, p. 71).

Perhaps the other side of the river is always greener (although German visitors to British council housing estates rarely go home thinking so). But German academics have always been inclined to apocalyptic views. Moreover, Kennedy's criticisms - in so far as they have some validity - seem to apply to West Berlin rather than 'West Germany'. In the early 1980s, the trade-union-sponsored housing organisation Neue Heimat initiated an urban renewal scheme in Berlin in a heavy-handed way, which provoked protests and 'occupations'; it subsequently faced scandals and financial difficulties. But the problems of Neue Heimat (discussed below) hardly justify a denunciation of 'the system'.

Whatever one's verdict on West German land and housing policies, they clearly need to be seen in relation to the general economic system of the Federal Republic. This system has been portrayed as 'free market' by critics ever since Lord Balogh denounced 'an iniquitous new German economic and social system' (Balogh, 1950). The belief, however, that West Germany practises laisser faire is demonstrably incorrect - in relation to land policy as well as other fields. There is far more intervention in the urban land market by public or quasi-public bodies than in either Reagan's USA or Thatcher's Britain. At the same time, there is far less centralised - and less discretionary - state power than in Britain. Power is widely dispersed, and exercised within a strict legal framework, changes in which require a large measure of 'consensus'. To understand the background to this situation requires some knowledge of German history.

HISTORICAL BACKGROUND

There are two themes in German history which have a direct bearing on urban land policy today. The history of Germany was, until only a century ago, the history of cities and small states. The tradition of local autonomy is therefore strong, and the Federal Republic is organised on a highly decentralised basis. National legislation lays down procedures, but substantive land use planning is in the hands of local authorities; appeals are possible to the Provincial (Land) Government, but only on whether the law has been complied with.

The second theme is that autocratic rule persisted into the 19th century and, 'The great aim of the movement against arbitrary power was, from the beginning, the establishment of the rule of law' (Hayek, 1960). The ideal of the Rechtsstaat was a system of laws which were of general applicability, thus curbing the discretionary power of rulers. There was to be a written constitution to lay down underlying principles, and a constitutional court to interpret and apply them. This ideal has never been wholly implemented in the field of town planning, since planning authorities find ways of retaining some flexibility and discretion. Nevertheless, the system which began to emerge in the 1870s was strongly based in law, and the experience of National Socialism inclined the Federal Republic still more to 'a government of laws rather than of men'.

Land Policy in German History

German urban history can be divided into:

1. The age of the citizen-ruled medieval cities, roughly 1000-1600 A.D.
2. The age of absolute rulers, roughly 1650-1800 A.D.
3. The 'laisser-faire' era of the industrial revolution, roughly 1800-1900 AD.
4. The 20th century, with its emphasis on communal town planning.

In the first two periods, there was considerable control of land transactions by the municipality or the local ruler. The idea that land was a good like any other, with which the owner could do as he wished, arose only in the 19th century, and soon gave way to the concept outlined under Eigentum (property ownership) in the Grosser Brockhaus encyclopedia.

'In the modern concept of ownership, limitations of legal, economic and social character are of great importance (e.g. town planning regulations). The modern social constitution is interlaced with such limitations, and in this way there has been a fundamental change from the concept of ownership in the classical-liberal period. But limitations on ownership rights do not destroy the essence of ownership. An intervention which removes the power of the owner in the economic management of the property is not a limitation of ownership, but a withdrawal of it'.

The 'Founding Years'

Current land policy has its origins in the urban explosion of the 1870s. At that time, urban population densities rose, since the new 4-5 storey blocks of rented apartments had higher densities than the older houses. At the same time, rents and land prices, in real terms, rose to higher levels than ever before - or since.

Two factors aggravated this increase in density and prices. Firstly, most people walked to work, while factories were situated centrally; moderately high densities were therefore inevitable. It was improved transport which chiselled apart the compact late-19th century city, and caused land prices to fall. Secondly, many towns were surrounded by fortifications and 'clear fire' zones which acted as a 'noose', limiting expansion and forcing up land prices. The army held on to these areas until a surprisingly late date: 1907 in the case of Cologne.

High-density housing, high land prices, and large gains from the sale of building land, led to a movement for the reform of housing and land policy (Dreier, 1968; Bullock and Read, 1985). The movement can be divided into three schools, all of which find echoes today; the Marxist/Georgian; the 'land reform' movement represented by Adolf Damaschke; and the 'mainstream' economists'. The first two, in particular, did not emphasise the factors mentioned above, but rather the influence of private ownership of building land and housing.

The first school was influenced by Marx and Henry George; it saw 'rent' and land prices as a tribute extorted by the 'land monopoly', which would vanish once universal public ownership ushered in an era of free land. Adolf Damaschke was less revolutionary. He distinguished between 'normal' land values and inflated 'speculative' values resulting from 'ill-conceived building plans, harmful building regulations, un-limited credit facilities, and unjustified privileges for un-developed land' (Damashke, 1920, p. 87). Damaschke gave pride of place to a tax on rises in land value. But he also advocated a 'greener' type of housing, municipal landbanking and leasehold disposal, and organisations which would build housing for sale or rent on a non-profit basis. His advocacy led directly to the setting up of 'homestead companies' and non-profit housing associations in the 1920s, as well as a tax on rises in property values.

Several 'mainstream' economists (A. Wagner, von Inama Sternegg, von Wieser) also examined the relation between land and housing problems. They stressed the peculiar (mono-polistic) characteristics of land, which distinguished it from other goods and justified more extensive state intervention. On the other hand, they rejected the abolition of private property and of a market economy in housing. They advo-cated the extension of city boundaries, the expansion of public transport systems, and active participation by the municipality in the land market, in order to exert counter-vailing market power and provide for the needs of 'social housing'.

Arthur Spiethoff, writing in the 1930s, when there had been a sharp fall in land prices, stressed that the building industry, and hence the demand for land, was characterised by strong cyclical movements - which he considered a justifi-cation for public 'land banking' (Spiethoff, 1934). He con-cluded that 'the housing problem' was not the inevitable product of 'economic rent' or a market economy in housing, but he attributed excessive density and unjustified gains by some landowners to defects in the land system. Although the profit motive had great virtues in the provision of housing, it served no purpose in the assembly of large 'greenfield' sites. Municipalities could easily acquire such land compulsorily at agricultural value, and sell it to builders. The provision of ample supplies of building land should be the basis of policy.

This analysis needs to be modified at a time when the main complaint about a market system is that it leads to 'urban sprawl' rather than excessive density. (One study indeed argues that 'speculation' and 'land hoarding' are beneficial (Risse, 1974).) However, the ideas of these 'defunct economists' on the role of the state in the urban land market contain much of lasting value, and influenced 'practical men'. Policies on municipal land policy and non-profit housing have displayed considerable continuity since the 1870s, in spite of wars, economic catastrophes and National Socialism.

THE ORIGINS OF MODERN LAND POLICY

The history of municipal land policy is particularly well documented for Cologne, and its experience is not untypical. In 1870, the city took over an ancient charitable foundation, which owned 3000 hectares of farmland. An 'estates department' (Leigenschaftsamt) was set up and, in order to gain financial benefit from the expected growth of the city, some remote land was sold, and land bought nearer the city. The estates department soon began to be used for housing and planning purposes.

> 'After the lost war of 1914 to 1918, the estates department faced a completely changed situation. Up till then, land transactions had been looked at primarily as a source of revenue. The emphasis now changed to land banking and the generous provision of green areas. At the same time land policy acquired a marked social slant. The then Mayor, Dr. Adenauer laid down three objectives: a generous provision of green areas; the concentration of industry in an industrial estate; the encouragement of suburban houses with gardens.' (Cologne City, 1974, p. 70).

In pursuit of these aims, the estates department became involved in the system of land reorganisation (Umlegung), which is Germany's most distinctive contribution to land policy. Umlegung evolved out of a procedure for consolidating farmland. Some parts of Germany did not undergo an 'enclosure movement', and the ending of the open field system left farmers with large numbers of small scattered strips of land. Consolidation through individual negotiation rarely succeeded. When, therefore, the farmers in a district agreed, a procedure known as Flurbereinigung was invoked. An indepen-

21

dent committee assessed the value of each farmer's land, and redrew the boundaries so as to produce consolidated holdings; the reorganisation was designed to ensure that the value of each farmer's holding was the same as at the beginning.

A comparable procedure was developed to deal with urban development on fragmented farmland, and later with redevelopment. An independent commission (Umlegungs-kommission) supervised the reorganisation of holdings into a pattern which allowed development but owners who did not wish to undertake development were given compensation in the form of money or land elsewhere.

The Prussian Land Reorganisation Act of 1902 (Lex Adickes), which laid down procedures for Umlegung, was applied in 1911 to the old 'clear fire' area around Cologne, which covered 360 hectares and contained 3000 holdings. The city drew up a plan for a ring of parks, together with roads and some housing. The Commission began work in 1921, and the 'park ring' was largely completed by 1939. There were similar developments in other cities. Thus, long before the setting up of 'town planning' departments in the 1960s, the principle of municipal participation in the urban land market was accepted, and has subsequently never been seriously challenged.

THE FEDERAL REPUBLIC

Land and housing policy in the Federal Republic has been dominated by the concept of soziale Marktwirtschaft (socially responsible market economy), which has come to be accepted by all three main political parties. The phrase - coined by, and best expounded by, Mueller-Armack - indicates an econ-omic system different from both a 'centrally planned' and a 'free market' economy. In its application to urban land and housing policy, soziale Marktwirtschaft has meant that a market system based on widespread ownership of real prop-erty - by individuals and housing associations - is considered desirable. But the state has an important role to play in alleviating poverty, ensuring that everyone has adequate housing, and tackling situations where the market does not work satisfactorily, e.g. in providing public transport, dealing with 'urban decay' and supervising large-scale land acquisition.

The Federal Republic has seen a marked change in political attitudes, as compared with the previous century of German history. In the framing of policy, great stress has been laid on 'consensus', and the mere possession of a Parlia-mentary majority has not been considered a justification for pushing through controversial legislation. In the 1980s, this consensus has become a little frayed, but has by no means broken down.

The 'social market economy' has not been without its challenges. The latest has come from the Greens, some of whom reject private ownership and a market economy in any form. But the more acceptable 'green' ideas are being absorbed into the philosophy of the 'social market economy'; a development of this sort was, in fact, foreseen by Mueller-Armack himself (Mueller-Armack, 1976).

The Constitution

The Constitution (Grundgesetz or 'Basic Law') of the Federal Republic (Article 14) lays down general principles which apply to real estate.

'(1) The right of property ownership is guaranteed. Its content and limitations are to be determined by law.

(2) Ownership involves obligation. The exercise of private ownership must also serve the public interest.

(3) Expropriation is permitted only in the public interest. It can be carried out only in accordance with a law which lays down the form and principles of compensation. The amount of compensation is to be based on a balanced assessment of the interests of the parties affected and of the public. In the case of dispute, the amount of compensation is to be decided by the courts.'

These principles are far from being an unqualified defence of the rights of the property owner. They have been used by the Federal Constitutional Court to rule that real property need not be treated in exactly the same way as movable property; that in cases of conflict the public interest must prevail over the individual; and that compensation need not necessarily be based on 'market value'.

The Constitution also specifies the powers, and the sources of finance, of the Laender and the local authorities - of which, since a reorganisation in 1969, there have been 8,500. Decisions on land use and housing development are in the hands of local authorities, which possess a high degree of status, power and financial autonomy (Eversley, 1974). The growth of increasingly complex legislation in the 1960s and 1970s, however, slightly reduced their powers. The Christian Democrats, who returned to power in 1982, stressed the need for more local autonomy. Chancellor Kohl stated that;

'Local democracy is, in our view, an essential part of our federal constitution. - The local authorities need elbow-room. We must simplify the law and remove excessive regulation. I am thinking in particular of building and planning law'.

WEST GERMANY

Documentation of Ownership
Germany has a comprehensive Land Register. Every site has
to be entered in the Register (Grundbuch) maintained by the
local authority. The site is given a number taken from the
large-scale maps (Kataster) drawn up for the whole country,
which show every land-holding. For each site, a dossier is
kept, which contains the required information and copies of
relevant documents. The Register is open to the public and
has to be consulted whenever a sale is made. The information
in the Register comprises;

(a) Physical details, e.g. '5 Schillerstrasse. House and
garden. 715 sq.m.'
(b) The name of the owner, and the date when he
acquired the property.
(c) Legal restrictions; tenancy rights, rights of way
etc.
(d) Financial encumbrances; e.g. mortgages.

All sales of real estate have to be conducted through a
'registered conveyancer' (Notar) who has the statutory
responsibility of giving independent advice to both buyer and
seller.

Freehold and Leasehold Tenure
Most real property is owned 'freehold', but there is a long-
standing provision for leasehold tenure (Erbbaurecht). The
minimum period of the lease is 33 years, but 66 or 99 are
more usual. The 'ground rent' (Erbbauzins) is normally set at
4-6 per cent of market value, and nowadays incorporates
provisions for periodic reviews. At the end of the lease, a
new lease is agreed, or the land and building revert to the
ground landlord. In the case of reversion, however, the
ground landlord has to compensate the leaseholder for the
'residual value' of the building. Although these provisions are
available, however, little use has been made of the leasehold
system since the 1920s. As a history of land policy in Cologne
explains, when reviewing the sale of municipally-owned land,

'In the period 1919 to 1924, about 90 per cent was
disposed of by means of leases -. As the ground rent
could not be raised during the period of the lease, the
inflation (the hyper-inflation of 1923) caused severe
financial losses to the city. After negotiation with the
leaseholders, agreement was reached in most cases to
change the leasehold into a freehold. After 1924, leasehold was used in only a very few cases.' (Cologne,
1974, p. 71).

Leasehold never came back into favour, and rent reviews remained unusual until the 1970s. Perhaps an opportunity was missed in the reconstruction period. If the cities which sold land at prices which now seem ridicuously low had instead granted leases, with provision for rent reviews, they would have acquired a useful income.

The Pattern of Ownership

Although the Federal Republic is densely populated, 'urban' uses account for less than 10 per cent of its land area. The agricultural land area is divided up into (in 1984) 641,000 holdings over 2 hectares; most holdings are owner-occupied, although some are tenanted (Table 3.1). The landlords include local authorities who have acquired 'land banks'; nearly one fifth of the agricultural area, and half of the forest area, is owned by state or municipal governments. A study of 'Who owns the Federal Republic?' has arrived at estimates of the area owned by various categories of owner (Table 3.2). Since these figures lump together agricultural and urban land, they are in themselves not very meaningful. However, a comparison with the situation in 1937 illustrates a growth of approximately one third in the land owned by state bodies, and a five-fold increase in land owned by housing associations.

Within town boundaries, typically 30-40 per cent of the area is owned by the municipality, in the form of streets, parks, various buildings and 'land banks'. Municipalities do not own housing on a significant scale, although most have their 'own' charitable housing association, in which they are the majority shareholder. Most housing is owned by individuals, for their own use or for renting, or by charitable (gemeinnuetzig) housing associations.

Table 3.1:   The Main Categories of Land Use: Federal
Republic, 1984

	Per cent of total area
Agriculture	56.1
Forest	29.5
Roads, railways	4.7
Buildings and surroundings	4.3
Water	1.7
Recreation	0.5

Source: Bundesminister fuer Raumordnung, Bauwesen und
Staedtebau, Bonn.

Table 3.2:   The Owners of Land: Federal Republic

	Per cent of total
Federal Government	2.76
State Governments	10.79
Municipalities over 10,000	2.60
Municipalities under 10,000	10.60
Religious bodies	3.95
Other public bodies	1.10
Individuals, couples and trusts	64.71
Housing associations	0.83
Private corporations	2.26

Source: Duwendag and Epping, 1974

THE PHASES OF HOUSING AND LAND POLICY

Three stages can be distinguished in the history of housing and land policy in the Federal Republic.

a.   Post-War reconstruction, with the emphasis on the quantity of housing.
b.   The era of 'grand designs' (say, 1965 to 1973).
c.   The era of renewal, conservation and 'quality of life'.

The massive housing programme of the first phase was not based on 'public housing', but on the participation of large numbers of private landlords and non-profit organisations. Federal subsidies were provided for 'social housing', but were available to any individual or organisation willing to let housing at the prescribed rents. Rent controls on pre-War housing were relaxed and eventually withdrawn (except in West Berlin), and there was no control on the rents of new private rented housing. In consequence, large quantities of private rented housing, as well as of social housing, were built. In the 1950s and 1960s, nearly one third of all new housing (and half of all new rented housing) was 'social', and was built by individuals, as well as 'housing associations'; in recent years, the amount of new 'social housing' has fallen, and it has been built only by housing associations. The emphasis has shifted from 'social housing' to personal housing allowances (Wohngeld), available for any type of housing.

Reconstruction
The rebuilding of the German cities took place under provin-

cial legislation on building and planning. 'Reconstruction Acts' were passed by the various Laender, based on existing provincial laws, with additional powers of compulsory purchase as a last resort. The starting point was the existing pattern of ownership. A building plan was drawn up, and discussions were entered into with the owners, to enable them, wherever possible, to rebuild within the plan. This often involved Umlegung and municipal buying and selling of land.

The Rebuilding of Cologne
The role of municipal land policy in the reconstruction phase is particularly well documented for Cologne. In 1946, the city council approved a policy of buying as much land as possible, on a voluntary basis, while selling land to anyone wishing to build housing. The legal basis for reconstruction was provided by the 1950 Reconstruction Act of North-Rhine-Westphalia and, in the same year, the city appointed a commission to undertake Umlegung. The commission delegated detailed work to a full-time executive who could call on the Estates Department and other municipal departments. A report on '100 Years of Urban Land Policy in Cologne' comments;

> 'This form of organisation has proved its value. Citizens affected by municipal planning do not suspect an independent commission of serving municipal financial interests in the way that they would suspect a department of the city administration' (Cologne, 1974, p. 83).

In the 'old town', which had been completely destroyed, considerable reorganisation was needed; some sites were only as big as one room. A problem elsewhere was that the provision of roads, parks etc. displaced tenants paying low rents, who would be unlikely to obtain similar cheap accommodation. The city sought to solve this problem by monetary compensation, but also by organising the construction, through the city's non-profit housing enterprise, of subsidised new housing. As the report comments acerbically, 'Ways and means were founded of resolving the resettlement problem, not through 'social plans' (see below), but through the exercise of sympathy and ingenuity' (p. 93).

Between 1954 and 1974, the reorganisation process involved 3178 sites, with an area of 798 hectares. In 471 cases there were objections, but in only 47 cases were these carried as far as an appeal to the courts. The average length of a reorganisation process was three years. Since the 1950s, the land bought and sold by the city has accounted for between 30 and 50 per cent of all transactions. Turnover peaked at around 600 hectares p.a. in the early 1960s, and has since fallen to around 200 hectares p.a.

27

WEST GERMANY

Reconstruction: An Assessment

The rebuilding of the German cities was for the most part architecturally conservative, consisting mainly of low-rise blocks, of conventional construction (Wedepohl, 1961). This rebuilding was treated condescendingly by 'modern' architects at the time, and a recent critic maintains that, 'It is almost impossible to make up for the lost chance in rebuilding residential areas in Germany after the War' (Kennedy, 1984, p. 73). A more balanced assessment, by a former Director of the Essen Estates Department, who was 'Left of centre' on land policy, is that;

> 'The rebuilding of the destroyed city districts was tackled with different objectives and different methods in the various cities of the Federal Republic. In the achievement of this great task, the political decision over the extent of public intervention in land ownership played a central role. The least successful rebuilding was where it followed the old site boundaries and building lines too closely. But alongside unsatisfactory solutions there are examples where the needs of urban improvement were taken into account. ... Green areas were incorporated in blocks which previously lacked light and air, while the needs of traffic were met by parking areas and garages' (Bonczek, 1971, p. 51).

The 1960s and After

In the late 1960s, a new generation of architects drew up plans for 'villes radieuses'. Many were fortunately not implemented - such as the plan by Neue Heimat to replace the entire district of St Georg in Hamburg with a single pyramid-like building 50 storeys high. The tower blocks which were built have often proved unpopular - although they do not display the decay and vandalism of some of their British or American counterparts.

This phase gave way in the early 1970s to one in which it was recognised that there was no longer a need for vast numbers of new dwellings; the emphasis shifted to the renewal of existing urban areas, and small-scale development. The Germans also began to discover the charms of the 'house and garden'; at the present time over half of the much reduced total of housing starts are of 'houses' (usually in terraces). Houses are nearly always built for owner-occupiers in the first instance, but the construction of adjacent houses, one to occupy and one to let, is not uncommon. Although owner-occupancy is gaining in popularity, there is still a viable and respectable private rented sector, which caters for the entire socio-economic spectrum. There is security of tenure, and a form of rent arbitration.

Table 3.3:   The Owners of Housing: Federal Republic

	Thousand dwellings	%
Privately rented	7,865	33.9
Rented from housing associations	3,259	14.0
Owner-occupied	9,321	40.1
Other*	2,778	12.0
TOTAL**	23,223	100.0

* Housing provided by employers; dwellings without a kitchen.
** Excluding holiday homes and old people's homes.

Source: Bundesministerium fuer Raumordung, Bauwesen und
          Staedtebau, Bonn

FUTURE NEEDS FOR HOUSING LAND

Since 1950, the total built-up area has grown faster than the total population, and has shown only a slight flattening-off since 1974, when population growth virtually ceased (Bundesminister, 1983, p. 43). The Ministry forecasts that, over the period to 2000, the area needed for housing will nevertheless continue to increase, for two main reasons.

(a) Although the population will fall by 2.4 mn, the number of households will increase by 1.4 mn.
(b) The floor area per person, which rose from 14 sq. m in 1950 to 34 sq. m in 1981, is expected to continue rising.

If more compact methods of building are adopted, this will reduce the demand for housing land, and the latest figures in fact show a slight fall in the area taken annually for new construction. It is forecast that, in the 1990s, the figure will fall by anything from 20 to 50 per cent. A comparison of projected demand with the areas zoned for housing suggests that, except in the cases of Munich and Stuttgart, sufficient land is available to meet the growth in demand. The 1986 'Land Report' also reviews the changing attitudes to land use. It has become widely accepted that the emphasis should be on Innenentwicklung - a new term which covers infilling, the reclamation of derelict land, the renovation and improvement of older housing, and environmental improvement (cf. Stadtsanierung, which has come to be applied to the thoroughgoing improvement of small areas). But although 'inner development' can make a significant contribution to meeting the demand for housing, some suburban development will, in the Ministry's view, continue to be needed.

The loss of agricultural land ('two large farms every day') has caused concern in some quarters. Similarly, 'urban sprawl' has been attacked. ('The German landscape dies' is a characteristically apocalyptic title of a 'Spiegel Book'.) But most foreign observers give high marks to the German landscape. German housing has always been relatively frugal in its use of land, and 'scattered' development has never been allowed. And although the old self-contained rural life has died, West Germany (especially in the South) has been more successful than many countries in creating a viable 'rural' economy. The movement of light industry into small towns and villages has helped to maintain population in country areas. It is common for a smallholding to be run on a part-time basis, with one or more members of the family working in a nearby town.

Finally, it is noteworthy that the 1986 Land Report, produced under a 'conservative' Government, is in parts distinctly 'green'.

'The realisation that the natural resource 'land' is limited, and performs ecological functions even within towns, which are endangered by a continuing extension of the built-up area, is relatively new. An important town planning function will in future be to reconcile the demand for new building land with the preservation of the natural functions of the soil'. (Bundesminister, 1986, p. 62).

MUNICIPAL LAND POLICY

Land policy in Germany embraces taxation affecting land use and town planning, as well as land policy in the narrow sense, i.e. public intervention in the land market. The latter is primarily a matter for municipalities, although some provincial governments also maintain important land banks. It is generally accepted that municipal and provincial governments should pursue 'active' land policies, in the sense of buying land in advance of development. One argument put forward in the 'Land Report' is that such policies lead to lower land prices (Bundesminister, 1986). The same document argues elsewhere that they do not have much direct effect on prices but can have beneficial effects on;

(a) the supply of land for 'disadvantaged' groups
(b) the utilisation of unused land
(c) a generally careful use of land.

Municipal land policies comprise:

    (a)   Umlegung.
    (b)   Land banking.
    (c)   Other actions under the Federal Building Act and the Urban Development Assistance Act.
    (d)   Participation in the land market using civil law procedures, i.e. the power to buy and sell in the same way as a private corporation.

Of (a) and (b), it need only be said that the long-established policy of Umlegung is still important, and that land banks are still being replenished; between 1969 and 1980 municipalities devoted 3-5 per cent of their capital budgets to the acquisition of land for which no immediate development was planned (Bundesminister, 1986, p. 20). Landbanking is still 'profitable', but the returns have fallen sharply since the 1960s. Local authorities are increasingly tending to act as intermediaries immediately prior to development, or to rely on other methods for influencing the land market.

The Federal Building Act

The two legislative pillars of town planning and land policy are the Bundesbaugesetz (Federal Building Act) of 1960, and the Staedtebaufoerderungsgestz (Urban Development Assistance Act) of 1971. The two Acts have been amended several times, and have now been consolidated into a single Act. Work on the Federal Building Act began in 1950, with the publication, as is usual, of a draft Bill, on which the comments of professional bodies and interested parties were invited. The Constitutional Court was also consulted, and it agreed only in 1959 that the Federal Government could legislate on town planning. After a great deal of examination and revision, the Act came into force in 1960. Its main provisions concern plans and powers to buttress them, and the provision of infrastructure (Erschliessung). Municipalities have to produce a land use plan indicating the type of development permitted in an area (Flaechennuetzungsplan). They may have to produce, when it is 'necessary for the requirements of town planning', a 'building plan' (Bebauungsplan), giving detailed plans of buildings, parks, areas with special environmental regulations etc.

The plans, according to the Act, should 'serve the social and cultural needs of the population' (which does not mean much because it could mean anything). Perhaps more meaningful is the provision that an extension of real estate ownership among the population should be encouraged. In the current legislation, which incorporates provisions pioneered in the Urban Development Assistance Act, procedures are laid down for public participation, and for a 'social plan'. The local authority is required to take account of any negative consequences for the local population (both property-owners and tenants) stemming from the 'building plan', and to investigate means of alleviating them.

There is no right of appeal against the provisions of a properly devised building plan. There is, however, a requirement that the plan should 'balance public and private interests' which the courts have increasingly taken as a justification for reviewing plans; several plans have been challenged by citizens in the courts, and declared illegal.

In the undeveloped parts of an area covered by a building plan, the municipality has a limited right of pre-emptive purchase (Vorkaufsrecht). The municipality may intervene in a planned sale, and require that it is sold the land, at the arranged price, or in some circumstances at an independently assessed 'market price'. This right can be exercised, however, only when it 'serves the public interest'; the local authority has to state the use to which the land will be put, and the claim can be tested in the courts. The local authority also possesses the right to forbid demolition, or to order an owner to build or modernise. In practice, however, local authorities rely on negotiation rather than on compulsion. There is a general power to initiate Umlegung, but compulsory purchase (Enteignung) can be invoked only where some public purpose can be achieved in no other way. The price to be paid is 'market value', ignoring 'hope value'.

The municipality is responsible for roads and paths, parks, play areas etc., and has to bear at least 10 per cent of the cost; it may decide to bear a higher percentage. The remainer is covered by an 'infrastructure charge' (Erschliessungsbeitrag) levied on the owners.

Civil Law Procedures

A municipality can also participate in the land market in a 'private' capacity. It can purchase land from the original owners when a building plan has been drawn up, and dispose of it to developers; if necessary, on concessionary terms. Alternatively, it can persuade the owners to grant 'easements' which give it a say in the selection of purchasers. To facilitate these land transactions it may, if it wishes, make use of a public agency. There are several such agencies; the best-known were originally known as Heimstaetten (homestead companies) but are now generally known as 'development agencies' (Landesentwicklungsgesellschaften). They are organised on a Laender basis, but operate autonomously, undertaking land deals and development for local authorities or housing associations.

LAND TAXATION

The history of land taxation before 1939 is instructive - and little known, even in Germany (Hallett, 1977, pp. 141ff). A tax on realised rises in land value was introduced in 1911. After 1918, with the fall in land prices, it ceased to bring in any revenue, and was replaced by the Land Acquisition Tax in 1936. In the same year, a price freeze was imposed by the National Socialist Government. The general price freeze began to break down after 1945, and was ended in 1948. Land prices, however, remained, in theory, frozen until 1960 - although in practice a black market developed. The Federal Building Act (in its eventual form) did not embody any provisions for the taxation of rises in land value. Its aims, according to the Federal Housing Minister of the time were,

> 'To end the freeze and to incorporate the land market in the social market economy; to simultaneously institute measures which will work against land-extortionists and ensure that a market emerges in which building land is offered at reasonable prices'.

The idea was to create greater 'transparency' in the urban land market, and thereby encourage greater efficiency and stability. 'Transparency' was to be achieved by setting up 'valuation commissions' which would publish statistics of the prices of land sold in development areas.

In the discussion of the Act, a 'betterment levy' (Planungswertausgleich) was envisaged, but it was not incorporated, after an expert committee had rejected it on practical grounds and the Constitutional Court had ruled that it would require a constitutional amendment. One provision for land taxation was, however, included in the Act - a penally high Grundsteuer (see below) when land zoned for building was not developed within a tight time limit. In practice, this provision proved inequitable, and was subsequently withdrawn.

Various land taxation proposals have been put forward over the years, but they have been criticised and rejected. As a number of studies have pointed out, taxes on 'development value' face considerable problems of valuation, and (at high rates of tax) can deter land sales (Risse, 1974; Felde, 1955; Hansen, 1955). A tax on site value or unrealised gains can cause inequity, or encourage undesirable development. Nevertheless, it might well have been possible to have devised a scheme which would have cut off some of the 'fat' of unearned increment without cutting into the muscle of 'incentive'. The general tax code also makes little provision for the taxation of rises in land values. Provided that an owner has held a piece of real estate for more than two years, and does not deal in real estate (or agriculture) as a business, no tax is levied on its sale. There is no tax on capital gains, but there is a wealth tax, of 0.7 per cent. 'Wealth' includes real property, the basis being the Einheitswert (rateable value, or assessed value) of housing, business premises or undeveloped land. These assessments are still based on a 1964 valuation, and so are well below current values.

One unfortunate consequence of the tax system is that some sellers of agricultural land can make large, untaxed, windfall gains; this has periodically been a source of criticism. However, a substantial proportion of this land is owned by public bodies. An often-cited study (in Der Staedetag, 11/1970) calculated that the gains on agricultural land sales in the period 1960-70 amounted to DM 50 bn. But 85 per cent of the land in the sample was owned by public or charitable bodies.

There is also an important exception to the freedom of land sales from taxation. Persons engaged in agriculture or forestry are charged Income Tax on land sales, unless the proceeds are reinvested in land. The tax rate in such a case can be 56 per cent (the maximum Income Tax rate), which has even been criticised as a deterrent to sales.

The arrangements for owner-occupied housing have recently been changed. Up till 1987, there was a tax on the 'imputed rent' of an owner-occupied dwelling, with an off-setting allowance for mortgage interest. The tax treatment was particularly favourable if a dwelling in a two-dwelling building (or a 'granny flat') was let. The tax on imputed rent, and the mortgage interest relief, has now been abolished, but anyone buying a house (new or old) can deduct up to DM 15,000 p.a. from tax for eight years, 'once in a lifetime'. The abolition of tax on imputed rent is described as the 'private good solution', the argument being that one does not pay an annual tax on a refrigerator or a boat. The point that one also does not receive a tax allowance appears to be overlooked.

Two further taxes should be mentioned; the Grundsteuer (local property tax, or 'rates') and the Grunderwerbssteuer (land acquisition tax). The Grundsteuer is one of the taxes used to finance local authorities, and is levied separately on bare land and built-on land; it is based on the Einheitswert together with the 'poundage' set annually by the local authority. It now constitutes under 5 per cent of local authorities' receipts, having declined in importance because of the under-assessment of the Einheitswert. This under-assessment has three further consequences;

(a) Suburban property is undervalued in relation to 'city' property.

(b) The tax on bare land has become negligible, which makes 'hoarding' costless.

(c) Real estate is (via Wealth Tax and Capital Transfer Tax) taxed less than financial assets.

The Land Acquisition Tax was originally levied at a rate of 7 per cent on all real property. This penalised owner-occupiers who moved house, and various exemptions were introduced; they have now been abolished and a flat rate of 2 per cent introduced.

Land Taxation: A Criticism

The main criticism of the current arrangements for land taxation is that the use of an outdated Einheitswert has produced fiscal distortions which arguably tend to raise the price of land, and provide excessive concessions for owners. Most economists favour a revaluation of the Einheitswert. Most politicians have opposed it, but there are now signs of a change, emanating from the bottom of the federal structure. The Federal Government plans (1987) to cut Income Tax, which would reduce the income of the states and communes. In response, the Government of Lower Saxony has proposed an increase in the Einheitswert. The influential Bund der Steuerzahler (Taxpayers' Association) has also come out in favour of such a policy, partly on grounds of equity. It is too soon to say whether these proposals will overcome the strong resistance to them.

Land Prices

There is a range of land types, and prices, from 'pure' agricultural land, through land with 'hope value' to 'raw' building land zoned for housing, but with no services, to 'land ready for building'. The rise in price from the first variety to the last can be twenty or thirty times, although much of the difference between 'raw land' and 'land ready for building' is accounted for by the cost of services (Bonczek, 1960). Official price statistics are published (in the Statistisches Jahrbuch) for 'raw land' and 'land ready for building'.

When the average figures are deflated by building costs and average incomes, a picture emerges of an upward trend since 1962, with considerable cyclical fluctuation (Figure 3.1). There was a rise to 1968 and then a fall to a temporary low in 1975, during a temporary glut of housing: in relation to income, land prices were lower in 1975 than in 1962. There followed a recovery to 1983, since when there has been an absolute fall of some 12 per cent, and an even greater fall in real terms. There has also been a fall in rents and property prices in many cities. The price of 'houses' has fallen by 17 per cent since 1981, while the price of flats has remained

Fig. 3.1:   Land ready for building. Average land prices in
relation to construction costs and average GNP

```
·········· Construction costs
——— Average GNP
1962 = 100
```

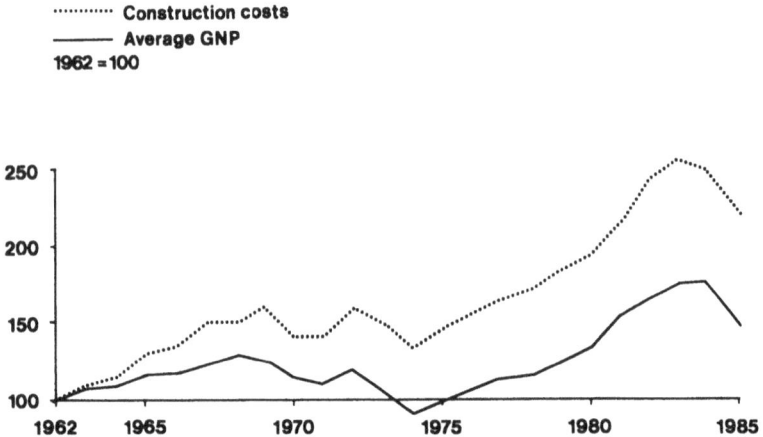

constant. The rents of 'free' dwellings in large cities fell by
7 per cent in 1985.

In real terms, land prices in the mid-1970s were below
the pre-1914 levels, the rise since 1962 being a 'catching up'
after the depressed levels of the 1930s. Between 1975 and
1983, land prices rose sharply in real terms. Since 1983,
however, a combination of factors – the balance in the housing market, the fall in the inflation rate (leading to less
interest in 'inflation proof' investments like land) and the
falling population – have had a dampening effect on land
prices.

The average prices in 1985 (first three quarters) for
'raw land' and 'land ready for building' were respectively DM
46 and DM 111 per square metre. These average figures are,
however, misleading. Prices varied enormously according to
the size of settlement. Using the 1983 figures, the range was
from DM 38 in settlements of less than 2000 to DM 333 in
cities of over half a million. Even between large cities, there
were big differences, mainly between the North and the more
booming South – from DM 64 in Luebeck to DM 632 in Munich.
There are also differences between (some) suburban areas,
where land prices are still rising, and (some) city areas,
where they are falling (Figs 3.1, 3.2).

Fig. 3.2: Land prices in various cities (land ready for building) 1970-85

DM per m²

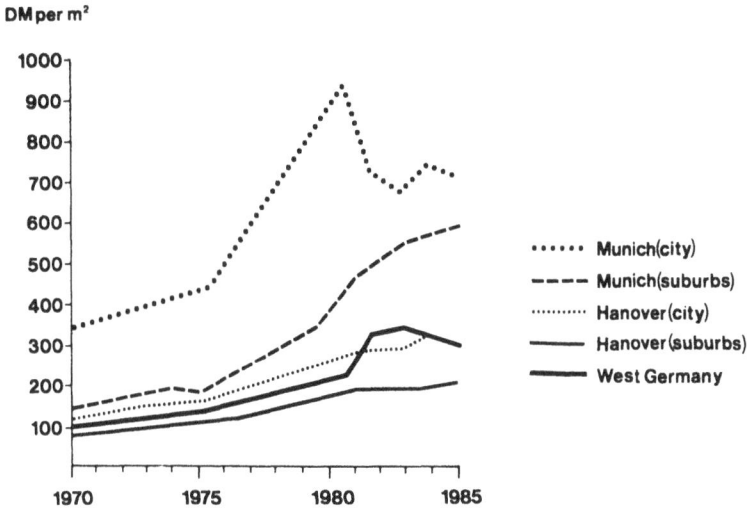

Land Prices and Housing
Critics of the prevailing system have in recent years argued
that land prices often account for 40 per cent of the price of
houses, thus pricing ownership out of the reach of many
people. A somewhat different slant on the subject is provided
by a Ministerial study covering the period 1975 to 1982
(Bundesminister, 1983). This concentrated on Stuttgart,
which together with Munich has the highest land prices in the
country, and Osnabruck - a fairly typical city. Land price
ranged from around DM 900 per sq. metre in Stuttgart city to
around DM 100 in the Osnabruck fringe. Housing was divided
into three groups; flats; 'compact low-rise', mainly terraces;
and detached houses. The cost of land (ready for building)
as a percentage of total costs for three types of housing is
given in Table 3.4.

These percentages vary less than land costs because, in
the cheaper areas, more extensive (and expensive) housing
forms were adopted. The average land area per dwelling was
around 100 sq. m. for flats, 200 sq. m. for compact low-rise,
and around 400 sq. m. for detached houses, but rising to
over 900. The report points out that land prices thus serve a
purpose in ensuring the economical use of scarce land, and
that a fall in land prices would not necessarily open homeownership to wider social groups, but might rather raise the

37

Table 3.4: The Percentage of Land Costs in Total Housing Costs, 1975-82

	Flats	Compact low-rise	Detached
	%	%	%
Stuttgart city	23.4	28.2	28.5
Stuttgart fringe	16.9	23.6	25.1
Osnabruck city	12.3	24.7	23.9
Osnabruck fringe	–	13.0	16.3

Source: Bundesminister, 1983

living standards of middle- and high-income groups. Its general conclusion is that measures directly designed to dampen land prices would not have any marked effect on the provision of housing, but could be only one element in a general review of housing policy.

SOCIALIST AND CONSERVATIVE CRITIQUES

Demands for the taxation of rises in land value built up during the period of rising land prices in the 1960s. After 1968, the whole 'system' was challenged by the student radicals, who argued that the allocation of land and housing should be carried out solely through a democratic political process, 'for use rather than profit'. Property rights would be replaced by a Nutzungsrecht ('right to use'). Occupiers would be entitled to use a dwelling for a limited period of time but would not have the right to transfer it to someone else. Such a system would seem to mean that, in the end, virtually everyone would be a tenant of the state, and all new housing would be built by public bodies.

This interpretation of Nutzungsrecht was certainly advocated by the far Left of the SPD, but other members envisaged merely a form of leasehold. A policy statement passed by the SPD in 1975 was ambiguous. It declared that 'ownership' should be replaced by either 'user rights' or an extended form of leasehold tenure, 'the choice depending on practicability and political feasibility'. The SPD/FDP Government did not follow either path. In 1980, however, it introduced an amendment to the Federal Building Act which provided for, in effect, the compulsory purchase of land at existing use value. With the accession of the CSU/FDP Government in 1982, this amendment lapsed.

The idea of substituting 'user right' for ownership has recently been revived by the Greens - or at least one section

Fig. 3.3: Land prices in various cities and GNP per capita
1970-85

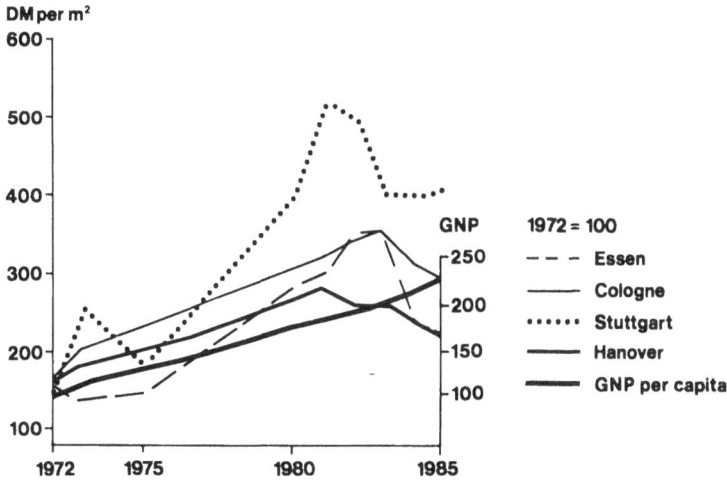

of this very diverse movement. The Fundis (fundamentalists) have argued that the problems of housing and ecology cannot be resolved as long as real property remains under the 'unrestricted control of private owners' and that all land and housing should be taken over by the (local) state.

Criticism of some aspects of the land system have also come from the CDU. One programme for reforming 'social housing' on 'market orientated' lines mentioned the land market (Biedenkopf, 1978). It attributed high land prices to a combination of unduly restrictive development policies by local authorities, and excessive fiscal privileges for owners. Urban land problems were less acute in the Federal Republic than in countries like Britain or France, because there were no monster cities like London or Paris. None of the conurbations exceeded about 2 million, and there was no reason why land should not be made available to meet demand. Local authorities, however, sometimes failed to release enough land for development because of misguided ideas about the virtues of high population densities; ulterior motives (such as the desire of SPD strongholds to retain population); or the costs of providing infrastructure. The report proposed that the 'infrastructure charge' be increased from a maximum of 90 per cent to 100 per cent. It also criticised the failure to tax rises in land value (without making specific proposals) and the use of an artificially low Einheitswert. Similar proposals are often put forward with the extravagance which the subject seems to encourage. One rather surprising example is an interview in Der Spiegel (10 Dec. 1984) by the President of the Federal Constitutional Court. The state had, he maintained, given away 'hundreds of billions of Marks' through not adequately taxing the enormous fortunes embodied in real property. These fortunes in the hands of the rich were a weapon against fellow men. The state did not remove the axe from the hand of the murderer, but even gilded its handle, as the judge put it, in terms reminiscent of Henry George.

The Property Owners Federation argued in reply that property owners were not all rich; about half the population owned real property of some type (Zentralverband, 1985). Who was going to pay the extra billions of marks a year which were to be raised by a higher level of Grundsteuer? Careful studies had reached the unanimous conclusion that the simplistic idea of 'taxing away all economic rent' would in practice cause inequities, and throttle the land market.

The philosophical debate on landownership and taxation thus continues, although most practical debate is concentrated in the 'middle ground', and many criticisms could be met by simply carrying out a new valuation for the Einheitswert. The most important changes in recent years, however, have been concerned with the renovation of older neighbourhoods.

THE URBAN DEVELOPMENT ASSISTANCE ACT

In the 1960s, it became apparent that there were areas of German towns which had not been destroyed, and were badly in need of renewal. They included both the medieval cores of some of the smaller towns and considerable areas of working-class housing built from the 1860s to 1914. The renovation of the small remaining 'medieval' areas is more 'the preservation of ancient monuments' than housing policy, but the renovation of the late 19th century housing, consisting mainly of walk-up blocks of flats of 4-5 storeys, can often be justified on economic grounds. Both types of renovation are covered, however, by the German concept of 'urban renewal', and considerable attention has been devoted to renovating medieval buildings and their surroundings in such a way that people are prepared to live and work in them (Bahrdt, 1973).

The 'Urban Development Assistance Act' (Staedtebaufoerderungsgesetz) was first drafted in 1960, but was passed in a considerably altered form only in 1971. During this period, there was a swing of the pendulum from 'comprehensive renewal' to a – sometimes exaggerated – emphasis on renovation and conservation. The Act was applied in a spirit of retaining buildings whenever possible. As the Act was passed by a 'socialist-liberal' coalition government – after having been initiated by a 'conservative-liberal' coalition – it also reflected the agitation in the 1960s for a recoupment by the community of the sharp rises in land values during this period. It contained provisions for a 'betterment levy' of the kind which had been proposed for the Federal Building Act, but not in fact adopted.

The Act deals with two types of policy;

(a) 'urban renewal' (Sanierungsmassnahmen)
(b) 'urban development' (Entwicklungsmassnahmen).

'Urban renewal' (or 'bringing back to health') applies to older urban areas which have ceased to provide satisfactory living and working conditions. 'Urban development' applies to the development of partially built-up areas, e.g. the expansion of ex-urban 'villages'.

The Act lays emphasis on 'participation', and stipulates that the interests of all those affected - owners and tenants and the general public - must be taken into account. The first step is to undertake a survey, which has to include an assessment of the existing social situation and the likely consequences of a 'renewal' policy, with particular reference to people who are likely to suffer. The subsequent 'building plan' has to be accompanied by a 'social plan' which makes proposals on;

(a) compensation for those who lose property interests in the redevelopment,
(b) financial assistance to tenants facing higher rents,
(c) 'social housing' and old people's housing in the new development.

The Act thus reflects a concern with the effect on people adversely affected by renewal, which was a reaction against an earlier style of purely physical renewal.

The 'renewal area' is then formally designated by the local authority. Thereafter, all sites in the area are subject to the provisions of the Act. The Building Plan and the Social Plan are then finalised, after individual discussions with all persons who would be affected. This consultation can be undertaken by the local authority itself, or by an independent, non-profit, agency. Even when an agency (Sanierungstraeger) is employed, as has happened in two-thirds of all cases, the final responsibility for the plan rests with the local authority. There have, however, been complaints of local authorities 'losing control'.

Experience has shown that a familiar dilemma arises when a detailed plan is given the status of law. Since renewal is a long process - much longer than the five years or so originally envisaged - a detailed plan tends to be overtaken by events. A plan consisting of general objectives, which does not tie itself down to the dimensions of every building, is more likely to stand the test of time, but it is not strictly compatible with the terms of the Act.

The Act distinguishes between Ordnungsmassnahmen ('reorganisation', perhaps) and 'building'. 'Reorganisation' includes the re-drawing of sites, the provision of roads and other infrastructure, the movement of people and businesses, and any necessary demolition; all this is the responsibility of the municipality. On the other hand, rebuilding and modernisation is the responsibility of individual owners.

The most controversial provision of the Act is the provision for a 'betterment (and worsement) levy' (Ausgleichsbetraegen). There had been a widespread feeling that property owners should not benefit from the rise in values associated with the renewal, when this rise had been made possible only as a result of public expenditure. The Act lays down that the agency organising the renewal - with the assistance of an 'expert committee' of valuers - shall make an assessment of the rise in property values which results solely from the renewal scheme, and to which the owner has not contributed. This amount has to be paid by the owner to the municipality, although there is provision for conversion into a loan in case of hardship.

In the event, the Act was followed by the sharp drop in land prices after the speculative excesses of 1972-74; a cooler economic climate; and the realisation that many 'inner city' areas faced economic problems resulting from the 'drift to the suburbs'. Although substantial levies have been imposed on some individual owners, the levy has not proved to be a significant source of income for local authorities.

The tendency in renewal areas is for the size of sites to be increased (larger shops, housing 'complexes'), leading to a decline in the number of property owners. There are, however, ways (encouraged by the Act, and recognised by property law) in which this tendency can be countered; previous owners may acquire one flat in a block, or become part-owners of a new shop. Any land acquired by the local authority has to be sold off when it is no longer needed. When the renewal process has been completed, the designation of the area is formally withdrawn, and the special provisions of the Act cease to apply.

The Aftermath of the Act

The renewal, to a very high standard, of fairly small areas under the Act, has proved very expensive. (In 1978, the costs were estimated at DM 8.9 billion for 578 projects, with an average size of 14 hectares (Bundesminister, 1978).) It has been argued that the Act's original procedures - although justified in districts of particular architectural importance - are too perfectionistic. There has been a shift to more piecemeal modernisation with an emphasis on energy-saving

(Modernisierungs- und Energiesparungsgesetz, The Modernisation and Energy Conservation Act 1978), combined with
infrastructural improvements providing acceptable conditions
at moderate cost, rather than 'ideal' conditions at very high
cost.

Expenditure under the Act has nevertheless continued.
It was around DM 200 bn. in the early 1970s, although it
more than doubled in the 'special programme' of public works
adopted as an employment measure in 1975-79. The subsequent CDU-led Government, although pursuing a controversial policy of cutting public expenditure, has concentrated
on current, rather than capital, expenditure. Indeed, for
1986 and 1987, expenditure under the Act has been increased
to some DM 800 bn.

The second part of the Act, concerning 'urban development', has been used far less frequently. The need for
large-scale 'greenfield' developments has passed, and this
part of the Act has been omitted from the unified legislation.

The 1985 Amendment

As a result of experience with the Act, an amendment was
discussed and, with general consent, came into force in 1985.
This amendment does not alter the provisions for the 'normal'
process, but provides for an alternative, 'simplified' process,
which can be adopted by the local authority if the situation in
the relevant area makes it appropriate. (The Land government
has the power to disallow the alternative procedure if it
judges that renewal would be thereby impaired; the local
authority can appeal to the courts).

Under the alternative procedure, certain formalities are
unnecessary, and no 'betterment levy' is chargeable. But
when roads, parks etc. are extended or improved, the local
authority can levy a charge on local residents. The 'normal'
procedure is still available, but with the addition of a de
minimis clause for the levy. If the receipts are likely to bear
no relation to the cost of collection, the levy can be
scrapped.

The Consolidation of the Two Acts

Since the Federal Building Act and the Urban Development
Assistance Act differ in several points, and Germany prides
itself on its codified system of law, the consolidation of the
two Acts has been discussed ever since 1971. It also began to
be felt in the 1980s that - after a series of piecemeal amendments - the whole corpus of law on town planning, land
policy and building (Staedtebaurecht) needed to be reviewed.
One conference summed up a widespread view when it called
for,

    (a)   a simplification and clarification of the law,
    (b)   a reduction in the administrative obstacles to small-scale development,
    (c)   a strengthening of the position of local authorities in relation to upper tiers of government (Schuster, 1984).

A 'Building Law Book' (Bundesbaugesetzbuch), incorporating both Acts, was introduced in draft form in 1983, and a line-by-line examination was begun: it was passed with all-party approval in 1987. The new legislation includes the following changes. The objectives of 'environmental protection' and of using land 'economically and with consideration' are emphasised. The right of pre-emptive purchase is slightly restricted. Building plans can be 'looser', and exceptions granted more readily. The 'infrastructure charges' remain, but they have to be returned if the infrastructure has not been provided six years after payment. The 'normal' and 'simplified' procedures for urban renewal areas are both retained. The changes are thus modest, but the whole corpus of urban law has been carefully reviewed in the light of experience.

RETROSPECT AND PROSPECT

West German land and housing policy has achieved a relatively harmonious relationship between market mechanisms and public intervention. Although there is considerable concern to protect private property rights, local authorities - using legal instruments and institutions built up over a hundred years - participate actively in the land market. This participation buttresses their town planning role, and makes it easier to bring about land-use changes by negotiation and persuasion rather than compulsion.

A considerable degree of consensus - in this as in other fields - has prevailed, and on the whole still prevails, in the Federal Republic. The Marxist Left, which would like to see all land and housing state-owned, has acquired a following in the universities, but little elsewhere. The New Right - which rejects virtually any role for the state in the land and housing market - has been even less influential. The emphasis on consensus and constitutionality is evident in the legislative process. German law-making resembles 'the mills of God'; the two main town planning Acts took ten years between drafting and passing. As a result, most snags are discovered before, rather than after, an Act is in force.

The Federal Republic has faced two major urban challenges. The first was to rebuild the destroyed cities, and provide housing for millions of refugees. It did so in a way which - in spite of some failings - is one of the undocumented triumphs in the history of city-building. In this achievement, municipal land banks and Umlegung played a role, but the pattern of ownership was also important. Housing was built by large numbers of individuals and housing associations. Subsidised and unsubsidised housing, tenanted and owner-occupied housing, were intermingled. This mixture was encouraged by;

(a) the subsidy system under which the provision of social housing was not confined to municipalities, or even housing associations,

(b) the concern with private property rights, which inhibited the use of large-scale compulsory purchase.

There were three beneficial consequences. Firstly, financial and human resources were enlisted, which would not otherwise have been available. Secondly, the fact that most landlords were responsible for only a small number of dwellings encouraged effective management and maintenance. Thirdly, the generally small scale of property ownership 'parcels,' and the intermingling of differing tenures, worked against the concentration of poverty, social problems and dereliction, which can initiate a cumulative downward spiral. A British geographer (who criticises the lack of British-style compulsory purchase, council housing and rent control) points out, after a survey in Cologne;

'... the striking degree of diversity within German inner cities, not only from district to district, but also at the narrowest local level. Quite commonly, the full spectrum of inner-city environments and housing conditions, ranging from the poorest to those which pose little or nothing in the way of residential problems, can be seen compressed together within the confines of one small locality' (Wild, 1983).

The main housing management problems have arisen in connection with one, untypically large, organisation - Neue Heimat. This organisation, sponsored by the Trade Union Federation (DGB), performed an invaluable function in the earlier period, but later became too administratively top-heavy, and acquired a touch of megalomania. It bought too much expensive land; built too many blocks of flats of the type which used to win architectural prizes; and undertook ventures in the commercial property market, and overseas, which lost money. The outcome was a massive accumulation of debts - and corruption scandals. The DGB, after struggling to turn the company round, attempted in 1986 to rid itself of the problem by handing over Neue Heimat's 190,000 dwellings - and its debts - to an obscure Berlin industrialist for a

nominal DM 1. This devious exercise was widely criticised as
an abdication of the DGB's responsibilities, and was thwarted
by the creditor banks. Partly as a result of Neue Heimat's
debts, the DGB is now sellling off a bank and an insurance
company which it owns. It is likely that Neue Heimat will
eventually be split up into regional groups owned by con-
sortia of housing associations, banks and other local
organisations. It is a sad end to the participation by the DGB
in the 'non-profit' housing movement. But it will by no means
be the end of that movement.

There is now a wide range of types of housing available,
both for rent and sale. Except perhaps in Stuttgart and
Munich, there is no 'housing shortage' in the sense that
people cannot find housing of what is generally considered an
acceptable standard. The 'occupations' of housing in 1980-82
are now a historical footnote. They may well have reflected a
yearning for a simpler. and more communal life-style (and as
such may be significant) but their transience suggests that
they did not, as some commentators maintained, reflect a
housing shortage.

However, housing is not cheap and, until 1983, it may
have been becoming dearer. According to the official statistics, the percentage of disposable income spent on housing
in 'middle-income families with two parents and two children'
rose from 11.3 per cent in 1973 to 12.9 per cent in 1983. The
figure for 'pensioners and recipients of social security benefits on low incomes' rose from 16.9 to 19.5 per cent. The
other side of the coin is the extremely high standard of
building, and the marked improvement in housing conditions
over this period. Since 1983, there has been a downturn in
average housing costs. Average figures can, however, conceal
important minority problems. Meeting housing commitments can
become difficult for people who are unemployed, and there are
now 750,000 people who have been unemployed for over a
year. However, these problems of (relative) poverty -
increasingly accentuated by family breakdown - are a matter
of employment and welfare policy, rather than housing or land
policy.

In the post-War period, the whole emphasis - justifiably
enough - was on housing supply, not on the taxation of
'unearned increment'. It might, however, have been possible
to have struck a somewhat better balance between equity and
efficiency. At the present time, there is a strong case for at
least conducting a new valuation survey for the Einheitswert
- which would not involve any changes in the tax system.

The second challenge faced by the Federal Republic was
to cope with inner-city decline, and to tame 'modern architecture', modern commerce, and the motor car, which between
them threatened to make cities 'unlivable'. If the achievements
of 'human town planning' have so far been less than triumph-
ant, they have nevertheless been considerable. The 'inner

city' problem has, however, been exacerbated in the 1980s by serious unemployment, and the presence of an unintegrated body of Turkish 'guest workers'. When the Turks move into a district, or a block of flats, the Germans tend to move out. This is a serious problem, on which little progress has been made.

During the remainder of this century, the influence of a falling population will be increasingly felt. The problem is no longer that of providing enough housing in total, and there will be no more large 'greenfield' projects. The emphasis will be on incremental development, of an 'environmentally friendly' kind. Older districts will be modernised and the tower blocks humanised, as far as possible; most new housing will consist of terraced houses and low-rise blocks of apartments for 1-2 person households, including the growing number of old people. Owner-occupancy will rise, perhaps to 50 per cent, but a viable private and non-profit rented sector will probably remain. Municipalities will continue to participate actively in the land market, both 'privately' and using the methods available in the 'Federal Building Law Book'. In view of the balanced housing market, the prospect of stable land prices, and the absence of 'time bombs' in the form of concentrations of severe housing deterioration, it seems unlikely (barring political cataclysms) that any dramatic changes in policy will be called for.

REFERENCES AND FURTHER READING

Bahrdt, H.P. (1973) Humaner Staedtebau, Munich

Balogh, T. (1950) Germany; An Example of 'Planning' by the 'Free' Price Mechanism, Oxford

Biedenkopf, K.M. and Miegel, M. (1978) Wohnungsbau am Wendepunkt, Stuttgart

Bonczek, W. (1960) Bodenwirtschaft und Bodenordnung im Staedtebau, Heft 9, Sonderhefte der Zeitschrift fuer Vermessungswesen, Stuttgart

Bonczek, W. (1971) 'Zur Reform des staedtischen Bodenrechts', in W. Ernst and W. Bonczek (eds.), Hannover

Brede, H., Dietrich, B. and Kohaupt, H. (1970) Politische Oekonomie des Bodens und Wohnungsfrage, Surkamp, Frankfurt. (A Marxist interpretation)

Bullock, N. and Read, J. (1985) The Movement for Housing Reform in Germany and France, 1840-1914, Cambridge

Bundesminister fuer Raumordnung, Bauwesen und Staedtebau, Baulandbericht (1983 and 1986) Auswirkungen unterschiedlicher Bodenpreise auf den Wohnungsbau, 1983 Schriftenreihe 'Staedtebauliche Forschung'; (1978) Erfahrung der Gemeinden mit dem Staedtebaufoerderungsgesetz, Schriftenreihe 'Stadtentwicklung', Bonn

Cologne City (1974) 100 Jahre stadtkoelnisches Vermessungs-
und Liegenschaftswesen, Cologne
Damaschke, A. (1920) Die Bodenreform, Jena
Dreier, W. (1968) Raumordnung als Bodeneigentums- und
Bodennutzungsreform, Cologne
Duwendag, D. and Epping, G. (1974) Wem gehoert der Boden
in der Bundesrepublik Deutschland? Bonn
Eversley, D. (1974) 'Britain and Germany: Local Government
in Perspective' in R. Rose (ed.), The Management of
Urban Change in Britain and Germany, London
Felde, H.W. vom (1955) Die volkswirtschaftliche Problematik
der Erfassung von Wertsteigerungen des Bodens,
Cologne
Frank, K. (1980) Handlexikon fuer Bauherren, Hauskaeufer,
Haus- und Wohnungseigentuemer, Goldmann, Bonn
Hallett, G. (1977) Housing and Land Policies in West Germany
and Britain, London
Hansen, J.R. (1955) Der Planungswertausgleich, Frankfurt
Hayek, F.A. (1960) The Constitution of Liberty, London
Kennedy, D. (1984) 'West Germany' in Martin Wynn (ed.),
Housing in Europe, Croom Helm, London
Mueller-Armack, A. (1976) 'Die zweite Phase der Sozialen
Marktwirtschaft' in Wirtschaftsordung und Wirtschafts-
politik, Stuttgart
Pergande, H-G. and Pergande, J. (1973) 'Die Gesetzgebung
auf dem Gebiete der Wohnungswesen und des
Staedebaues' in 50 Jahre Deutsche Bau- und Bodenbank
Aktiengesellschaft, Frankfurt
Pfannschmidt, M. (1972) Vergessene Faktor Boden, Bonn
Risse, W.K. (1974) Grundzuege einer Theorie des Bauboden-
marktes, Bonn
Schuster, F. (ed.) (1984) Neues Staedtebaurecht, Konrad-
Adenauer-Stiftung
Speithoff, A. (1934) Boden und Wohnung in der Marktwirt-
schaft, Jena
Wedepohl, E. (ed.) (1961) Deutscher Staedtebau nach 1945,
Deutsche Akademie fuer Staedebau und Landesplannung
Wild, T. (1983) 'Residential Environments in West German
Inner Cities' in Trevor Wild (ed.), Urban and Rural
Change in West Germany, Croom Helm, London
Zentralverband der Deutschen Haus-, Wohnungs und Grund-
eigentuemer e.V. (1985) Jahrbuch 1985, Duesseldorf

Chapter Four

THE NETHERLANDS

Barrie Needham

INTRODUCTION

It is a remarkable thing about The Netherlands that it has
the biggest 'land problem' of any Western European country
(land is scarcest there) but that this results in fewer social
and political problems (land speculation, unearned capital
gains on land, distortions of urban development and
redevelopment) than in most other countries. The Dutch may
not be able to make land plentiful, but they have prevented
the land scarcity from creating subsidiary problems.

For the purposes of this book the possible problems
concern housing: high land costs in housing; or difficulties
with making land available which delay housebuilding; or the
fact that housing cannot be provided where it is desired
because the land there is not released. In this way also, the
land scarcity has not been allowed to create problems. In The
Netherlands, decisions are taken publicly about the desirable
production and renovation of housing, by type of housing,
price, location, amount and timing. Those decisions are
supposed to be in line with general housing and town planning policy statements. And land is regarded as a necessary
and neutral input for carrying out the decisions (comparable
with the building industry). Land policy must be in the
service of, and an instrument for, housing and town planning
policy; it must not influence the content of those policy fields
(see e.g. Delfgauw, 1934; v.d. Hoff, 1976; Kruijt and
Needham, 1980).

A central feature of Dutch land policy is that the instruments have been created by the national government but are
applied almost entirely by the local (municipal) government.
And although the provincial and national governments can
exert much influence over that local application, in practice,
land policy is local policy. As such, local governments are
free to use it as a way of realising their local policies (for
housing, town planning, economic development, etc.). It is
land as an instrument in the service of local policies that has

received the attention, far more than the ideological issues arising from land ownership and land prices, issues which in other countries have kept land policy as a topic for national government. The irony is that the Dutch, by carrying out for many years now an active land policy locally, have so altered the land market that the phenomena, which in other countries have led to an ideological land policy nationally, have now largely disappeared.

That affects the structure of this chapter. First we SET THE SCENE by describing:

- demographic and social changes and the effect this has on the scarcity of land
- recent changes in land prices, the land cost in housing, and the share of 'development gain' in housing, these being a reflection of the effects of land scarcity
- the actors involved in the supply of and the demand for land and the effect this has on the structure of the land market
- some features of housing policy and the housing market.

Secondly we list briefly the LAND POLICY INSTRU-MENTS which have been created by central government for the use of local government. It will be seen that this list contains nothing spectacular, nothing which could explain the singular success of the Dutch in making of their land policy a practical instrument for housing and town planning instead of a political football. The explanation of that success is that the municipalities add to the powers reserved for public bodies (as set out in the list) the very active use of those powers available to private bodies: the municipalities buy and sell land just as any private body may.

Thirdly we describe the MUNICIPALITY's LAND POLICY, that is how those instruments (public and private) are used for realising local policies for housing and town planning. We treat:

- the difference between an active and a passive local land policy
- greenfield development
- urban renewal
- financing this policy
- the consequences for land prices.

Fourthly we discuss the FINANCIAL PROBLEMS which the municipalities are now experiencing, as a result of working in that way.

Finally we draw some CONCLUSIONS which should be applicable to other countries also.

SETTING THE SCENE

Demographic and Social Changes

The population of The Netherlands in 1983 was 14,362,000: in 1953 it was 10,493,000: in those 30 years the population grew by 37%. Comparing that with the rate in some other West European countries (e.g. Belgium 12%, Denmark 17%, France 28%, West Germany 25%, UK 9%) we see that the pressure on land has been increasing very rapidly in The Netherlands.

The population density in 1983 was 352 people per sq. km: the corresponding figures for some other West European countries are: Belgium 323, Denmark 119, France 100, West Germany 247, UK 228. Measured in this way, land is very scarce in The Netherlands. (Source of all the above data, Statistical Yearbook, UN.)

The pressure on land is affected not only by the number of people living within the country, but also by what they do. If those activities result in more space per person for housing, for work, for recreation, for transport, that also raises the demand for land. This is indicated by the following figures. Between 1956 and 1983, when the population grew by 31%, the number of households (including 1-person households) grew by 80% (Statistisch Zakboek, Centraal Bureau voor de Statistiek/CBS). Households are becoming smaller as children leave home earlier, as old people live independently for longer, as couples divorce or maintain two households.

In 1958, the total area of land in urban use (defined for these purposes as land used for traffic, recreation, industry and commerce, other built-up areas, building land) was 237,600 ha. or 6.6% of all land (excluding the Ijssel Lake and Dutch parts of the North Sea): in 1983 when the population had grown by 27%, the area of land in urban use had grown by 115% and was 13.7% of all land (Statistical Yearbook of The Netherlands, CBS).

As is well known, the Dutch have been reclaiming land, partly in response to the land scarcity - 312,600 ha. since 1542 (Schulz, 1983; Meijer, 1979 gives 692,000 ha. since 1200 AD!). However, it was argued by van de Zwaag (1977) that this was no longer a sensible response to land scarcity. A dyke has been built around 41,000 ha. of water (the Markerwaard) in the Ijssel Lake, but the area has not yet been pumped dry: van de Zwaag (op. cit.) calculated that if there was no change in the rate of conversion of rural to urban land, an area equal to the net addition made by the Markerwaard would be fully occupied within 3 to 5 years! It follows that if the Dutch are to solve the problem of land scarcity, this must be by using less land per person. Perhaps that is one of the reasons why the present government has refused to finance the draining of the Markerwaard, leaving it to private initiatives if they find it profitable!

Land Prices and Land Costs

In Table 4.1 we present land prices (guilders per sq. metre), between 1965 and 1982, for land sold within the agricultural sector, for land bought for building upon, and for land sold fully serviced for building (for the latter two categories, only the land bought and sold by municipalities: the reason is explained below). It will be seen that agricultural land prices rose rapidly between 1965 and 1978, since when they have fallen slightly; that the price paid for agricultural land for building upon has followed a similar pattern; that the price of building land sold serviced by municipalities has risen steadily and much more rapidly than the other two sets of prices. The probable explanation is that inflation and demand for building land have pushed up agricultural land prices, but not dramatically; and that the rise in the price of serviced land is caused by great improvements in the standards of that servicing (see below for what "servicing" includes).

The big difference between the price of agricultural land and the price of serviced building land is not caused by development gains (i.e. the effect on land price of granting planning permission on agricultural land) but mainly by the servicing costs. These can be exceedingly high in a country with such poor soil conditions. Most land has to be drained before it can be built upon and sometimes the level has to be raised by a metre or more by pumping sand onto it. The land is sold fully serviced, with access roads, street lighting, parking spaces, planting, etc.

Calculations made for 1975/76 gave the following acquisition costs:

-   agricultural value of land   f 3 per sq. m.
-   acquisition costs of greenfield land
    (excluding compensation for buildings
    etc.)   f 7 per sq. m.
-   therefore, development gain   f 4 per sq. m.
-   total acquisition costs   f11 per sq. m.
-   therefore, costs of compensation etc.   f 4 per sq. m.

But it then cost around f 38 per sq. m. to service that land for housing use (what follows does not apply to industrial land), giving a total cost of f 49. Only 50% of land acquired is disposed of (the rest is used for roads, open space, etc.): selling 1 sq. m. of building land must, therefore, cover the costs of 2 sq. m. The average disposal price was, therefore, f 100 per sq. m. (Kruijt and Needham, 1980, app. 5). Those calculations have not been up-dated, but more recent information (see Table 4.1 and NIROV, 1985) shows that the amounts have not changed much.

For a cheaper house for owner-occupation with all-in building costs of f 105,000 and a plot area of 160 sq. m., total land costs (f 140 per sq. m., somewhat higher than the

Table 4.1: Land Prices (Guilders per sq. m.) 1965/82

Year	Agricultural land (trans- actions within agriculture sector)#	Price paid by municipalities when buying land for urban development (excl. prices for buildings on the land)	Price paid to munici- palities for fully serviced building land	
			For industry	For all other uses
1965	0.85	4.00*	13.20	20.30
1966	-	5.60*	9.60	25.00
1967	-	6.50*	11.50	29.60
1968	-	6.60*	14.20	32.40
1969	-	7.70*	20.60	37.90
1970	0.89	9.00*	20.70	38.60
1971	-	8.90*	19.40	44.50
1972	-	9.10*	25.10	44.70
1973	1.37	10.30*	22.88	54.20
1974	1.61	-	-	-
1975	1.98	-	-	-
1976	3.01	6.70	35.20	73.60
1977	4.20	8.70	37.70	80.30
1978	4.35	8.50	41.60	90.90
1979	4.04	10.10	45.70	111.20
1980	3.39	10.90	52.10	114.90
1981	-	10.50	40.90	125.70
1982	-	8.20	100.20	139.50

Notes: #: for agricultural land the prices are for 1965/66, for 1966/67 etc.
*: for 1965 to 1973 these prices include price for buildings on the land.
-: prices not known.

Sources: Maandstatistiek bouwnijverheid, CBS.
Statistiek van overdrachten en verpachtingen van landbouwgronden, CBS.

average) were 17.5% of total costs, development gain only 1% (calculation made for 1975/76). Similar calculations made for a typical private house in S.E. England in 1977 gave land costs there as 23.4% of total costs and development gain as 16.1%! (Kruijt and Needham, 1980, app. 5). For rented housing ('w-w-w' see below), land costs were around 17% of total costs (DGVH, 1985).

It is these figures which enable us to say: the problem of unearned capital gains consequent on the granting of planning permission, a problem which has dominated the land policy of so many countries, is of much smaller proportions in The Netherlands. Obviously, a more than doubling in price (from f 3 to f 7 per sq. m.) when land is taken from rural to urban use is exceedingly interesting for the land owner, and attention to this has been important in <u>national</u> politics (via the question of the price for compulsory purchase, see below). But the effect on house prices is negligible and the wish of land owners to realise the development value "floating" over their land has not been allowed to influence decisions about the location of urban development. The great scarcity of land is not reflected in that way.

Actors in the Land Market

The land market can be considered with respect to both urban development and urban redevelopment (renewal), and the way in which both those processes take place will be described later. For the moment it is sufficient to note that, until the last 10 years or so, there was hardly any urban redevelopment: the Dutch cities are comparatively young and have been well maintained. So most land transactions (apart from within agriculture and forestry) have been for urban development - land taken from rural into urban use, in particular to provide all those extra houses which the Dutch people wanted. Between 1958 and 1983, the stock of dwellings increased by 121% whereas the population increased by 37%: the absolute increase in the dwelling stock was 2,833,000; the number of new dwellings built in that period was 3,200,000; it follows that most of the new dwellings must have been built on greenfield sites, not on sites cleared of old housing (Statistisch Zakboek, CBS).

The process is simple to describe. Most rural land is supplied by the farmers or other land owners (sometimes speculators: the development gains, if small, are still worth having) to the municipality. (Note: there is no 'land banking', the municipalities do not acquire land far in advance of need because it is so much cheaper then: it is not, in fact, very much cheaper then!)

The municipalities then service the land and supply it to the developers - for housing, industry, shopping, etc. Obviously not all land thus acquired is disposed of: that needed for roads, planting, etc. (in housing areas, as much as 50% of all land, see above) is retained by the local authority.

Table 4.2 illustrates this. We see that in the 5 years 1978-82, municipalities acquired 8480 ha. of land for building upon, and disposed of 7980 ha. of building land (obviously many of those disposals would be of land acquired before

1978). In that same period, 2420 ha. of building land were supplied by others. Of the 10,400 ha. in total disposed of for building in that period, 77% were supplied by the municipalities. If land for industrial uses is excluded, 7860 ha. of building land were disposed of in that period, of which 79% by municipalities. It is apparent that the latter dominate the market for building land.

Development may take place only on land within a designated plan area and it has happened that some of the land within that area, and needed for building, had been bought in advance by property developers. They could refuse to sell to the municipality, which could not use compulsory purchase if the developer could show that he was able to provide the development exactly as shown in the plan. The advantage to the developer of supplying not only the development but also the land is that it strengthens his bargaining position vis-à-vis the municipality. Compulsory purchase powers now cover this kind of situation.

In other words, when buying land for urban development the municipalities are monopsonists, when selling serviced building land they are monopolists. Municipalities can be regarded as the main 'producers' of building land. They do this in order to be able better to realise their housing and town planning aims, in ways described later. Another topic for later is the consequences of this type of land market for land prices.

Table 4.2: Land Bought and Sold by Municipalities and Others (ha.)

Year	Land acquired by municipalities for building upon	Land supplied by municipalities for building upon		Land supplied by others for building upon	
		All	Excl. ind. land	All	Excl. ind. land
1978	1448	2069	1520	690	533
1979	1783	1794	1256	513	312
1980	1990	1552	1201	490	281
1981	1642	1269	1116	325	257
1982	1617	1295	1109	402	277
1978-82	8480	7980	6200	2420	1660

Source: Maandstatistiek bouwnijverheid, CBS

One final point about the land market. It will be apparent that a municipality so heavily involved in buying and selling land needs a strong 'land department'. This is the grondbedrijf, that is so influential in the local land market.

The Housing Market

In order to appreciate the significance for housing of the Netherlands' peculiar land market one needs to know that most new housing is built by housing associations and by private developers. (The municipalities have in the past built and managed many rented dwellings but have now disposed of most of their stock to housing associations, and the municipalities now build very few new dwellings a year: a few houses are built 'one-off' by owner-occupiers. Note that the 'social housing' is not public housing.)

New housing is divided into the following 'financial categories':

- woningwetwoningen
  rented dwellings built under the Housing Act, the finance for which is lent by central government. The developer receives an annual subsidy which keeps down the rents. These w-w-w are the nearest equivalent to Britain's council housing.
- premie-woningen
  these are both for rent and for sale, and central government guarantees the private loans which the developer has to take for the construction. On the rented dwellings the developer receives an annual subsidy so that rents can be reduced: on the dwellings for sale the buyer receives a subsidy towards the mortgage costs for a limited number of years.
- vrije sector
  central government is not involved in the financing of these dwellings, which can be for both rent and sale. No direct subsidy is given (although tenants might be entitled to rent rebates and owner-occupiers can get tax relief on mortgage payments).

The housing associations build the subsidised rented housing (w-w-w and premie-huur) which they subsequently manage, also the housing for sale with a temporary subsidy (premie-koop). The private developers build housing for rent (both with the lighter subsidy - premie-huur - and unsubsidised) which they subsequently sell to a property company for letting, also housing for sale (both the subsidised - premie-koop - and unsubsidised).

The new housing built in 1983 had the following composition:

```
- rented          : for 1- and 2-person households    9,800
                  : Housing Act (w-w-w)               42,600
                  : premie-huur                       22,000
                  : other rented                       1,100
- owner-occupied  : premie-koop                       30,000
                  : unsubsidised                       5,600
- total                                              111,100
```

That housing was provided by the following agents:

```
- commissioned by central and local government       6,700
- commissioned by housing associations              49,000
- commissioned by institutional investors            8,900
- speculative builders                              33,300
- other private                                     13,200
- total                                            111,100
```
(Statistical Yearbook of The Netherlands, CBS)

This way in which housing is supplied has two important consequences:

-   the municipality, when disposing of housing land in a plan area, negotiates with a few housing associations and property developers. This can proceed quickly and professionally.
-   as so much of the housing (even that built by private developers) built receives a subsidy in one form or another (95% in 1983, see above), most of the housing cannot be started until financial permission has been granted. The subsidies come from central government and are channelled through the municipality. This strengthens greatly the position of the municipality when negotiating with the developers (both private and the housing associations).

It will be apparent that for urban renewal the ownership of the existing housing (especially in the older urban areas) is important too, not just who builds the new housing.

The composition of the housing stock by tenure is shown in Table 4.3 for the country as a whole. But in the older areas, privately rented dwellings are much more important, e.g. 52% of the housing stock in Amsterdam in 1982 (see Cijfers, 1982), and it is these which are so often in a poor state of repair. The municipality has powers to oblige the private landlord to repair his dwellings (aanschrijvingbeleid – notification policy), but this power is difficult to enforce and often the only solution is for the municipality itself to acquire the dwellings.

In the older areas, the other dwellings which so often need repair are those owned and occupied by poorer people. Improvement grants are available, but even so rehabilitation

Table 4.3: Composition of the Housing Stock, 1984

rented	2,926,000 (=44% of total)	
of which	from housing associations	51% of rented
	from local and central government	15% of rented
	from other non-profit organisations	7% of rented
	from private persons	16% of rented
	from institutional investors	11% of rented
owner occupied	2,252,000 (56% of total)	
total	5,178,000	

Source: Nationale Woningraad, 1985

to an adequate level is often financially impossible for the owner-occupier.

Thus it is that the municipality often ends up owning a significant part of an urban renewal area – of housing to be improved, of housing to be rebuilt, of land to be cleared (e.g. of factories) and used for housing. It is the housing associations which provide most of the new housing in those circumstances: private developers can find little to interest them financially.

Also important for urban renewal and the improvement of housing is the question of rent control and the quality of the housing stock. The present rent policy (huurbeleid) came into force in 1979 and has as its main aim to make rents comparable for comparable dwellings. This is especially important in a time of inflation, for the rents for new dwellings could grow seriously out of line with the rents for existing dwellings (set by historic costs). There is a national system whereby rents are set by reference to the amenity offered by the dwelling and whereby the amount by which the rents may rise annually is determined politically. That system, in combination with housing improvement subsidies, helps to keep the existing housing stock in reasonable repair: Table 4.4 presents some results of a recent survey of housing conditions.

One final point is necessary for understanding the relationship between a municipality's land policy and its housing policy. To build subsidised dwellings you need financial permission from central government (the subsidiser): but the construction of all dwellings (for sale and for rent) requires permission supplied by the central to the local government. Annually each municipality is informed of the quota (contingent) of dwellings (specified by type) that may

Table 4.4: Quality of the Housing Stock Built Before 1971

That proportion which is in moderate-to-poor structural
condition

		%
- socially rented	built before 1939	56
- socially rented	built before 1945	41
- privately rented	built before 1939	59
- privately rented	built before 1945	42
- owner-occupied	built before 1939	45
- owner-occupied	built before 1945	29

Source: Kwalitatief Woning-registratie, 1985, CBS

be built. That has two important consequences. The
municipality does not begin the expensive servicing of building land until it has received the necessary permission (or
the promise of it) from the centre. And the position of the
municipality in negotiations with the providers of housing is
strengthened still further: not only does the local government
supply the land and act as an agent for central government
in passing on subsidies, also it applies for and passes on the
necessary 'housing permissions'.

Until recently, the housing market in The Netherlands
has been tight, with long waiting lists and hardly any un-
occupied and difficult-to-let dwellings. That explains the
emphasis on building on greenfield sites (expanding the
housing stock) and the late emergence of urban renewal and
housing improvement. It explains also, in part, the squatter
movement which, especially in Amsterdam, has attracted
international attention. It should be added that squatters are
often motivated by ideology as well as by the difficulty of
finding accommodation: for although in most places the housing shortage has diminished in the last few years, and
substantial numbers of flats in less attractive housing estates
are now standing empty, squatters continue to take over, and
occupy, vacant premises (Ministry of Housing, Physical
Planning and Environment, 1984).

PUBLIC INSTRUMENTS OF LAND POLICY

Here we shall restrict ourselves to those instruments which
are available to public bodies only and, as we shall see, there
is nothing special about such instruments.

The Land-use Plan (Bestemmingsplan)

This plan is by far the most important instrument for town planning in The Netherlands, as it is the only type of spatial plan which is legally binding on both the citizen and public authorities (for an extended discussion in English of this type of plan, see Thomas et al., 1983). Municipalities are obliged to make a land-use plan for all land outside the built-up area, in order to control development there: such plans are used mainly for conserving, for preventing change. When municipalities want to promote change, then they usually make a land-use plan for that purpose also, for the area concerned. In this chapter - interested as we are in the relationship between policies for land, planning and housing - we concentrate on the use of land-use plans for promoting change: the following discussion is not in every respect applicable to plans made for rural areas.

For the land on which the development is to take place, the municipality makes a land-use plan. Legally, a valid land-use plan has two consequences for the power of the municipality to control the development. First, anyone wishing to build has to apply for a building permit: that is so on all land, but on land for which there is a valid land-use plan the permit may not be granted if the application does not conform to the plan. In that way, the land-use plan gives the municipality the power to control development passively, by refusing permissions, Secondly, having a valid land-use plan is one of the conditions on which a municipality can purchase land compulsorily (see below): there are other grounds for compulsory purchase, but having a land-use plan gives one of the easiest methods. The way in which a municipality uses its ownership of land to guide the execution of the development will be made clear later.

A valid land-use plan has one further significance besides the legal powers it gives to a municipality. We have seen that a municipality applies to central government for permission to allow houses to be built: that permission is not given if central government considers that the municipality cannot ensure that the sites on which to build the housing are available - for example, by having a valid land-use plan.

An obligatory part of a land-use plan is a FINANCIAL STATEMENT which shows how the costs incurred by the municipality are to be covered. The costs include:

-   the costs of making the plan itself (either by the municipality's own staff or consultants commissioned by the municipality). This can be covered by levying a standard charge of between 19 and 22% on total land servicing costs (Prins, 1985).

Whenever the municipality has acquired the land on which the development is to take place (and that is in the vast majority of cases) then the following have to be added:

- costs of acquiring the land and paying compensation
- costs of servicing the land. (These will usually include the costs of draining the land, sometimes even of raising its level - 'the low countries'!)
- putting in gas, water and electricity, foul and surface water drains, roads, footpaths, cycle tracks, parking spaces, street lighting, etc.)
- interest charges on the capital costs until these latter are recouped.

Very often, a large-scale development requires new infrastructure which benefits the whole town. Then there has to be added:

- a part of the costs of those 'supra-district facilities' (bovenwijkse voorzieningen). (Who is to pay the remainder of those costs is a matter for negotiation between municipality, province and central government.) For that part which is to be recovered from the plan itself there are a number of possibilities (see de Graaf, 1983, para. 1.7) - e.g. a charge of 10% of total land and servicing costs or f 3 per sq. m. land disposed of. In that way an 'urban development fund' can be built up and differences in such costs between various large-scale developments evened out.

The income can be of three sorts - from levies, from land disposals, from subsidies. If the municipality were to restrict itself to making the plan and issuing permits (i.e. no direct involvement in the land) its own costs would be low and could be recouped by levying charges on the developers. Where the municipality acts as the land developer (the usual case, see above), it can raise income by disposing of the land - either freehold or by building leases. If those two sources of income are inadequate, then there must be subsidies. Under stringent conditions and when town planning considerations make it preferable to build on more expensive locations, central government will pay to keep down the cost of housing land in expansion schemes - lokatie-subsidie - and there are subsidies available for urban renewal schemes. Otherwise the municipality itself must subsidise. The financial statement must show how the costs are to be covered. Losses are not acceptable; profits are not forbidden; but the expected aim is to do no more than cover the costs.

Compulsory Purchase Powers (Onteigening)
According to the Dutch constitution, land can be acquired compulsorily either by an Act of Parliament for each particu-

lar case or under the terms of an Act regulating compulsory purchase generally. It is this latter (the Compulsory Purchase Act - Onteigeningswet - revised in 1981) which now regulates the practice. Of particular interest is article 77, which provides for compulsory purchase in the interest of housing or planning policy:

- for implementing or maintaining a land-use plan
- for implementing a building plan
- for acquiring land for housing
- for acquiring and demolishing housing declared unfit for habitation.

Recently it has been established that, under certain restricted conditions, these powers can be used for acquiring housing in order to improve it (see below).

When a land-use plan is the basis on which compulsory purchase is applied for, the application is tested against:

- the planning and housing interests
- the public interest (in particular, could the existing owner carry out the planned development himself? And has the municipality already tried to acquire the land amicably?)
- the necessity
- the urgency
- the financial feasibility

The other admissible grounds for compulsory purchase (see above) involve procedures much longer than when a land-use plan forms the basis: this latter tends to be used most frequently, therefore. It was explained above that one of the most important functions of a land-use plan is that it makes compulsory purchase possible.

The price that must be paid in such a situation is specified as follows - the price that would arise by an assumed transaction in free negotiations between the compulsory purchaser and the compulsory purchased, both acting as reasonable people (art. 40.b). This is not existing use value, nor market value, but the value in the use for which the land is being purchased, determined in an 'objective' way (de Haan, 1981). Furthermore, although the new land use is allowed to influence the price, the necessary public works may not (art. 40.c). The result is what the Act calls the 'actual value' of the land.

But how can one give a meaning to such a metaphysical construction? In practice, the courts use precedence and tend to fix the compulsory purchase price at a multiple of existing use value. And this then determines the price for most amicable transactions also; there is little difference between the price for amicable and for compulsory purchase (Maandstatistiek Bouwnijverheid, CBS). One consequence is easy to predict: the land owner has little to gain by refusing to sell and forcing a compulsory-purchase process. Between 1979 and 1982, 99.94% of all land bought by municipalities for building upon was acquired amicably (Maandstatistiek van de Bouwnijverheid, CBS).

Also financially important is the compensation which must be paid, over and above the price for the land and buildings: the principle is (Art. 40) that the person compulsorily purchased is compensated fully for all the loss caused directly and necessarily by the taking of his property.

It is difficult to understand why the issue of the price by compulsory purchase has been so politically loaded: it was the immediate cause of the fall of the coalition government in 1977. Opinions were polarised around whether existing use value should be paid or 'actual' value: but as the figures given above show, the difference is not great (a factor of two) and has a negligible effect on the total cost of housing (see pp. 52-4). Obviously, for the person whose land is bought the difference is considerable: equally obviously, there are ideological issues at stake. The formula adopted is a political compromise: and in practice it does not need to be applied often (for 0.06% of all land acquired!).

The Law on Urban and Village Renewal (Wet op de Stads- en Dorpsvernieuwing, 1985)

This has given three new powers - an 'urban renewal plan' the 'regulations concerning amenity', and the 'right of pre-emptive purchase'. The urban renewal plan (stadsvernieuwingsplan) has the same status as a land-use plan (see above) and is subject to the same procedure. It has, however, much closer links with implementation in that to the purposes for which compulsory purchase can be used have been added modernising and replacing buildings (see 'compulsory purchase'). Further, the urban renewal plan has to be accompanied by an implementation plan. The procedures for designating an urban renewal plan are the same as those for a land-use plan.

The amenity regulations (leefmilieuverordening) have a duration of up to 10 years and can be used to combat deterioration in the living and working conditions in a given area. Certain functions can be forbidden, and that is an extra ground against which the application for a building permit can be tested. The applicant can be required to deposit a sum of money in his bank, to give more certainty that the building permit will be taken up. And the municipality can require private land-owners to permit temporary facilities (e.g. parking, play space) on their land. Such amenity regulations cannot be used for compulsory purchase, but they acquire legal status much more quickly than does a land-use plan.

The Right of Pre-emptive Purchase (Voorkeursrecht)

Under this power, the municipality has the first choice whenever someone wants to sell land or buildings: and the seller has the corresponding duty to offer his property to the municipality. This can have considerable advantages over the use of compulsory purchase: it can be applied much more quickly and to prevent speculation which might drive up prices. But it is a power which can be used only under precisely specified conditions: recently it has been restricted to land designated as such and included within an urban renewal plan (see above). If the owner of such land wants to sell, he must inform the municipality: the municipality must decide within two months whether to buy: then follow negotiations over the price which, if necessary, will be determined by a court of law.

The Tax on Certain Development Gains (Baatbeslasting)

If the municipality has carried out works which have raised the value of property (such as improving the accessibility, improving the water control) then it may levy a tax yearly for up to 30 years on the property concerned. This is equivalent to a tax on 'betterment proper' (Clarke, 1965). In practice, however, the legal and administrative complications are so great that this instrument is hardly ever used (v.d. Hoff, 1976).

It should be mentioned that municipalities levy an annual tax on property, 84% of all municipalities raising the tax on the value (rather than the size) of the property (Jacobs, 1984). But although this value is supposed to be revised every 5 years in line with market changes, the tax is too low to allow an effective transfer to the public purse of unearned capital gains.

Controls on Land Prices

By a law of 1981 (wet Agrarisch Grondverkeer: law on transactions in agricultural land), when agricultural land changed hands, the price was subject to controls, but only if it remained within agricultural use. If it was sold for urban use, that control could not be exercised. That was the only control on land prices, and the legislation was weakened in 1985 (Docter, 1987).

THE MUNICIPALITY'S LAND POLICY

All those instruments just mentioned are available to the municipality, but what is of at least equal significance is that the municipality supplements those public powers with the private powers available to everyone - the power to operate actively on the land market.

Local Land Policy, Active or Passive

In order to understand this better we make the distinction between a passive and an active local land policy. A municipality using the minimum public powers would, when it had made its plan, sit back and wait for people to apply for planning permission. This is the passive approach; others take the initiative; the power of the local authority consists of saying 'no' or 'yes' to such initiatives. Such an approach characterises land-use planning in Britain and many other West European countries.

The active approach is when the municipality, having made its plan, uses the power of the land-owner to help realise it. Obviously the first step is to become a land-owner, to acquire the land. The supplementary powers lie in being able to dispose of that land. You can choose to whom to dispose of it, and (within market constraints) at what price, also when. And you can dispose of it under conditions far more detailed and far-reaching than may be imposed when granting formal building or planning permission.

In The Netherlands, such powers of the land-owner are particularly wide, for it is not only restrictive covenants that may be imposed (as in Britain, specifying how the land may not be used) but also positive obligations (such as that the person buying the land must maintain the building up to a certain level, must contribute annually to the costs of communal facilities, etc.).

Disposal can be either freehold or leasehold. Disposal by leasehold increases still further the conditions which can be imposed on the use of the land: but it is interesting that it is legally possible, even when disposing freehold, to impose conditions which bind not only the first buyer but all subsequent buyers (kettingbedingen). A survey in 1981 of all municipalities with more than 25,000 inhabitants (this covered more than half of the national population) discovered that about 85% of municipalities disposed of land normally freehold: however, 80% of municipalities used leasehold disposals also. Thirty per cent used leasehold frequently and 12% used this exclusively (see de Jonge, 1984).

It is interesting that leasehold disposals have so far rarely been used positively as an instrument of housing and town planning policy. Leasehold gives the land owner (the municipality) the possibility of (re-)gaining possession of the land more quickly than by compulsory purchase, to impose and enforce conditions about maintenance etc., and when the leasehold expires to steer redevelopment in desired directions. Also, leasehold enables the municipality to capture (a part of) betterment. Instead of using those powers in that way, however, the biggest municipalities have in recent years tried to use the system as a source of revenue, linking the ground rents to the rate of inflation; the political outcry was so loud that those municipalities have modified their systems. It is to be hoped that they will use leasehold as a positive instrument for housing and planning rather than as a source of income.

It will be understood that a municipality which takes an active land policy is in a much better position to achieve its housing and town planning goals than is a municipality with a passive land policy. In The Netherlands, most municipalities pursue an active land policy. To the reasons which have led to this practice belongs the 'technical' fact of poor conditions: building land has had to be reclaimed from the sea or from lakes and swamps, it is badly drained or the water table is very high. There are therefore significant economies of scale in making land serviceable. Further, in the frequent fights against flooding it has become accepted that individual interests in land be subjugated to public interests. This helps to explain not only how widespread is the practice of public land development but how old it is: it was taken for granted in many of the larger municipalities around the turn of the century (Venverloo and Tromp, 1972), and van der Valk (1986) has recorded how the municipality of Amsterdam planned and directed its expansion in the second half of the 19th century by selling building concessions on the land it owned around the edges of the town.

The comparison with Britain is again instructive. There, most local authorities have a passive policy. It is the New Town Development Corporations which, in their operation on the land market, most nearly approach the Dutch practice. Nevertheless, it would be incorrect to generalise from this that Dutch authorities are more 'development oriented' than their English counterparts. With respect to the land market this is indeed so, but in other respects (e.g. managing rented housing, running schools) the Dutch municipalities are less involved, as Thomas et al. (1983, pp. 74-7) point out.

How this active local land policy works out in the practice of greenfield development and urban renewal will be described below, also how it is financed and the consequences for land prices. (For a further discussion in English, see Thomas et al., 1983, especially chap. 3.)

Greenfield Development
Remember that the municipality usually owns all or most of the land on which the development is to take place. The

authority then puts the services in - draining the land (lowering the water table or raising the land surface), putting in gas/water/electricity, building roads and parking spaces, providing playing areas, public open spaces, etc. To finance those works (usually exceedingly expensive) the municipality generally borrows the money on the security of the building land. The ground not used for roads and public service (about 50% of the plan area, see above) is disposed of for housing (mostly), also schools, a shopping centre, an old persons' home, etc. Disposal is to a developer - for housing to a commercial developer or a housing association; for shopping to a commercial developer; for schools, old people's housing etc. to a semi-private trust or foundation.

We restrict ourselves to housing, by far the largest user of the building land. The municipality chooses a developer or, for the larger expansion schemes, several developers, sometimes by open tender, sometime on the basis of 'it's your turn next', sometimes on the basis of experience with successful cooperation in the past. The land is transferred to the developer under a contract in which is stipulated not only the price of the land but also the types of housing to be built (to rent or buy, method of financing, size, etc.), the number to be built, the price/rent, the delivery date. In this way, the municipality can have detailed control over what is built.

Then comes the design and building stage. A smaller authority with a small expansion scheme usually enters into an agreement with one or two developers and lets them get on with the job. A larger authority divides a large expansion scheme into several projects, then for each project appoints a developer and forms a 'building team'. This building team contains a representative from the developer, the job-architect, an urban designer from the municipality, and the building contractor. Such a building team is responsible for the detailed design and for supervising the realisation of the project. In that case - it will be clear - the municipality is very closely involved in the building phase also.

Urban Renewal
Until the Urban and Village Renewal Act became law in 1985, the public powers of the municipality were the same as for greenfield development, and the private powers (to act in the land market) likewise. However, the circumstances surrounding urban renewal are different in many respects, and for the local authority the following are particularly important in practice:

-     the land is owned by very many more people, which makes acquiring land costly and time-consuming, and the time at which the necessary land will become available is difficult to predict

- the land is more expensive, with an urban instead of a rural value
- acquisition usually involves not only the land but buildings and works also, and this too increases costs greatly
- if carried out humanely, the process takes a long time, for very many residents and small businesses have to be provided for. The result is that interest costs on loans can mount frighteningly.

The result is that an active local land policy on the scale of that pursued on greenfield sites is financially impossible. Land acquisition has to be much more selective and many parts of the plan can be achieved only 'passively': i.e. by specifying what will be allowed, supplemented perhaps by the offer of grants.

The new law of 1985 gives the municipality more public powers (see above) and reduces administrative restrictions on the grants paid from central to local authorities, but it does not make it significantly easier for the municipality to pursue an active land policy in urban areas. (It is relatively easy, of course, for a local authority to paralyse the land market in an urban renewal area, by causing planning blight etc. But we would not like to describe that as an 'active local land policy'!)

Financing this Policy

As we saw above, for all land-use plans a financial statement has to be prepared. However, the accounts (as distinct from the budget) do not have to be made public, and the municipality often has an interest in not letting the financial realisation of its plans become known: especially if things turn out well, the municipality wants to transfer some of the profits to the reserves in some 'laundered' form. The result is that no statistics are available for the whole of the country, and an attempt to analyse the accounts for a few municipalities resulted in more frustration than enlightenment (SEO, 1979).

We must count ourselves fortunate, therefore, that we have the following figures for the municipality of Rotterdam. These show very clearly the financial differences between greenfield development and urban renewal (Table 4.5).

The Consequences for Land Prices

We said in the introduction that the continued application over many years of this type of local land policy had so altered the land market that significant development gains no longer arose, and later we presented some figures which substantiated this. But how are land prices set in these circumstances?

Table 4.5: Costs of Greenfield Development and Urban
Renewal

	Greenfield development	Urban renewal
	%	%
COSTS		
– land acquisition	9	55
– servicing	87	34
– other (incl. professional costs)	4	11
– total	100	100
INCOME		
– from disposals	72	15
– contributions from third parties	6	
– subsidies	22	85
– total	100	100

Source: de Graff (1984)

When the municipality buys land for urban development,
it must pay at least agricultural value for a voluntary sale.
For a compulsory sale, the municipality could – perhaps – use
its monopsonistic position to pay less: but this would be
socially unacceptable, and the price to be paid is set by the
law on compulsory purchase.

When the municipality sells (or leases) the land, serviced
for building, the market is a little 'freer', for often
municipalities are competing with each other to dispose of
land. The lowest disposal price is set by the financial
necessity of covering all the costs. The highest price is set
by the competition with other municipalities (in this respect
there is a kind of land market!); by the Ricardian fact that
land will not be sold for a price higher than its 'residual
value' (although the land market is not very free, the market
in many types of buildings is!); and by the principle that it
is not correct for a public authority to make heavy profits
out of land transactions (see also Needham, 1983). The result
has been, usually, disposal prices a little higher than cost
prices. In that case, it can be said that the municipality has
realised a (small) development gain. Where the market price –
or for certain subsidised uses a norm-price – is below the
cost price, then disposals will not cover costs and subsidies
have to be added to balance the books. This is often the case
in urban renewal areas (see above).

Where subsidies are not necessary, the total development
gain can be regarded as being divided between the seller of
the rural land (the farmer or land owner) and the seller of
the building land (the municipality). In both cases, a public

authority is closely involved and tempers pressures to raise land prices.

That discussion is, however, incomplete, for it takes no account of disposals to the various uses within the plan area: at what price is land sold for shops, for playing fields, for private housing, for rented social housing, etc. In more detail, then, the practice is as follows. Where it is possible to determine a market value for the intended use (e.g. for industry, shops), up to this value can be charged. For some social uses (e.g. libraries) a norm-price is stipulated. Subtracting the income from those sources from the total costs that must be covered gives the costs to be recovered from housing uses: those costs are divided between rented flats, rented housing, owner-occupied housing, terraced housing, semi-detached housing etc., according to the size of the plot and certain arbitrary indicators. It is, therefore, total costs which are covered by total disposals.

Four conclusions follow:

- this system can be used to cross-subsidise within a plan area, charging higher prices for commercial uses in order to reduce prices for 'social' uses
- the system is flexible, allowing profits to be made or, if the land prices would otherwise be higher than market prices or some norm, subsidies to be added as income
- the influence of the municipality on land prices is even greater and more detailed than might at first sight have appeared
- the municipality can use this power to influence prices as a policy instrument, for example charging low prices for industrial land as a way of attracting firms (see e.g. Needham, 1982; Needham, 1983).

This description of how land prices are set might be difficult to accept for someone familiar with the land markets in Britain and North America and the theories developed to understand them. So be it: in The Netherlands land prices are set very differently, mainly because of the very different structure of the land market there.

The method outlined above (whereby total costs are divided between the plots to be disposed of) is not obligatory (except for urban renewal plans) but is strongly recommended, and in one form or another it is followed by all municipalities. The method was first set out in 1968 (Grondkosten woningbouw) and has come under increasing criticism. In particular the arbitrary division of costs and the principle of 'covering the costs' leads to insufficient attention being paid to quality differences and the price/quality relationship, also to disposal prices which bear little relation to market value. The calls are growing louder for a method which takes more account of the fact that land, especially when developed, is a negotiable good and as such has a market value (Federatie "O"/RIGO, 1987).

CURRENT FINANCIAL PROBLEMS

There is, however, another aspect to the financial consequences of the Dutch local land policy. It is that active participation in the land market carries with it financial risks. You incur enormous debts which you hope to be able to recoup (or even better) by disposals. That there is a risk involved is so obvious that one wonders why it has received so little attention in The Netherlands. Presumably, inflation and the steadily rising land prices since the end of the Second World War have ensured that, providing the municipalities were careful, they could always cover their costs (with popular suspicion having it that they have done better than that, thank you!). It was like a business venture that could not fail.

The last few years have changed that, however, and drastically. Decisions to buy land have to be made far in advance, so when the economic recession set in, many municipalities had already committed themselves to acquiring land, based on financial calculations which assumed a steady demand for that land, rising land prices and low interest rates. The economic recession brought with it a stagnation in demand (especially for private housing: for ordinary single-family housing, if the average sales price was 100 in 1965 it had risen to 460 in 1978 but fallen to 396 in 1984 - Maandstatistiek Bouwnijverheid, CBS); stable land prices (see above); and higher interest rates. Even worse, many municipalities had reacted to the rising unemployment by buying land on to which industry would be attracted: but very many municipalities were competing for the same firms, most of which did not in any case materialise! It has been estimated that of all the land available on industrial estates, only one half is in use (Bak, 1985).

The problems were compounded by a change in national government policy, whereby it was decided to give 'housing permissions' to the cities and designated 'growth centres', rather than to the smaller towns, where previously the growth had been, and where the municipalities had already acquired land for further growth (see Giebels et al., 1985, p. 94).

That an active local land policy involves great financial risks soon became apparent. A number of studies have tried to quantify what those risks have led to.

The Province of Gelderland (Provincie Gelderland, 1983) estimated that the municipalities in that province had, in 1982, reserves of land for housing sufficient for 10 years, for industry sufficient for 10-18 years, for other uses sufficient for 16 years. The costs of acquiring and servicing that land had been f 1,100 m., on which the municipalities were paying interest charges of f 100 m. a year.

Another study, restricted to industrial land (Ike et al., 1984), found that at the beginning of 1983 municipalities had land reserves of 10,500 ha. (with a book value of f 8,600 m.), harbour boards had reserves of 2,400 ha. (book value f 1,600 m.), and private developers had reserves of 4,000 ha. (book value unknown).

A third study (Giebels et al., 1985) estimated that, at the beginning of 1983, municipalities owned 100,000 ha. land in total, of which one half (50,000 ha.) was intended for disposal (equivalent to the reserves, indicated by the first two studies). Of those 50,000 ha., 30,000 had already been serviced: they had a book value of f 5,000 m. The 20,000 ha. not yet serviced had a book value of f 3,000 m. The munici- palities estimated (on the basis of very favourable assump- tions!) that their capital loss - after disposals - on the above land would amount to f 2,500 m.

No one, however, has a solution for this problem. People can say what must be done in order to prevent it arising again, but what is to be done now? Someone must cover the huge losses. Probably the answer will be found in hard- fought negotiations between national, provincial and local governments.

Those financial consequences have planning consequences also. The advantage of an active local land policy, it was said above, is that with it a municipality can exert a strong influence on what is built. When municipalities are competing against each other to get rid of their land as quickly as possible, they are not likely to use their (private) powers as land-owners to stipulate extra planning conditions. The effectiveness of local land policy is, thereby, reduced.

CONCLUSIONS

The first conclusion is that, in a country where land is scarce, it is not inevitable that private land owners can profit from that scarcity by driving up prices, extracting develop- ment gains and - directly or indirectly - influencing the form of urban development. Nor are those the only problems that can arise: ideological objections to that private power can lead to discontinuity in public policies; the private search for development gains can distort urban development and the planning system, leading (the experience in other countries shows) to cases of bribery and corruption. In The Nether- lands the connection between land scarcity and the power of private land owners has largely been broken by the munici- palities playing, for very many years, a very active role on the local land market.

The main reason why municipalities pursue an active land policy is, of course, not that they wish to dominate the land market but that they wish to have more powers for realising their housing and town planning aims. The second conclusion is that this can be very successful. The high quality of the housing and of the built environment in The Netherlands is undoubtedly attributable to a very large extent to the fact that the municipalities are the suppliers of most of the building land, and that they use this position most positively and creatively; the land cost in housing is not high; housing is built when and where it is desired, not when and where the land market makes it possible; some 'social mix' is achieved within neighbourhoods by not disposing of land for huge one-class housing estates.

Those two conclusions could be of interest to other countries also, but then they must take careful note of this third conclusion: that following an active local land policy is financially very risky. The Dutch municipalities have realised this, very late, and their reactions to it are very interesting. For you hear nowhere the cry: we must withdraw from an active role on the land market and let others take the financial risks! An active local land policy is considered essential as an instrument of housing and town planning. But what has to change is the blind acceptance of the consequential financial risks. This will involve very much more sophisticated budgeting and a different attitude to making profits on land deals. In particular, where there is financial risk, the municipality must do what a businessman would, and budget for profits in order to cover unforeseen losses.

That suggestion was made by Giebels et al. (1985) in a report commissioned to investigate that general problem. The report considered the following also: that the freedom of municipalities to pursue a local land policy should be restricted by closer controls exerted by the province. That also has a more general purpose. When municipalities operate on the land market as the Dutch authorities do, they are strong in a sellers' market but weak in a buyers' market; moreover in that latter situation they can weaken each other further by mutual competition. When a public authority operates so actively on the land market, it must protect itself by co-operating with other public authorities, not in a conspiracy against private interests but to safeguard public money and public interests.

The last conclusion of general relevance is the need, financially, for continuity. Land policy can be a most effective instrument of housing and town planning policy; but it demands continuity and a long-term view. The main reason is that the procedures for acquiring land and making it available take so long (often land has to be bought 10 years before it can be disposed of, because the official procedures take so long): when land is disposed of leasehold, there is the addi-

tional fact that the returns are realised only over the very
long term. If land policy has to react to short-term fluctu-
ations in housing and planning policy - whether those
changes are made centrally, provincially or locally - then the
financial costs can be very high.

REFERENCES AND FURTHER READING

Bak, L. (1985) 'Collectieve Bedrijventerreinen', in Nederland,
    Bijdragen tot de Sociale Geografie en Planologie, no 12,
    V.U., Amsterdam
Cijfers (1982) 'Amsterdam', in Jaarboek 1982, gemeente
    Amsterdam
Clarke, P.H. (1965) 'Site Value Rating and the Recovery of
    Betterment', in P. Hall (ed.), Land Values, Sweet and
    Maxwell, London
Delfgauw, G.Th.J. (1934) 'De Grondpolitiek van de Gemeente
    Amsterdam', H.J. Paris, Amsterdam
DGVH (Directoraat-Generaal van de Volkshuisvesting) (1985)
    Jaarverslag 1984, Staatsuitgeverij, Den Haag
Docter (1987) 'Town and Country Planning', in European
    Environmental Yearbook, London
Federatie "O"/RIGO (1987) Uitgangsputen Hierziening Grond-
    kostenmethodiek, Amsterdam
Hoff, C.M. van den (1976) Aspecten van Grondbeleid, VUGA,
    Den Haag
Giebels, R., Koopmans, C.C., Moolhuizen, F. (1985) De
    Risico's voor Gemeenten op de Markt voor Bouwrijpe
    Grond, Sitchting voor Economisch Onderzoek, Amsterdam
Graaf, B. de (ed.) (1983) Grond en Ruimte, Samson?, VUGA,
    Den Haag
Graaf, B. de. (1984) 'De Exploitatie van Grond', lecture
    given at conference 'Inleiding in het Gemeentelijke
    Grondkostenbeleid', Amersfort
Grondkosten woningbouw, Vereniging van Nederlandse
    Gemeenten/Ministerie VRO (1968), Den Haag
Haan, P. de (1981) 'Grondpolitiek: een groen licht en een
    nieuw programma', in Nederlandse Juristenblad, 21 Feb.
Ike, P., Voogd, H., Zwieten, K. van (1984) Bedrijf-
    sterreinenplanning, Moeilijkheden en Mogelijkheden, T.H.
    Delft, Delft
Jacobs, A.G.A. (1984) 'De Onroerend-goedbelasting in
    Nederland 1983', in Openbare Uitgaven, vol. 16 no. 6
Jonge, J. de (1984) Gemeentelijke Gronduitgifte, Kluwer,
    Deventer
Kruijt, B. and Needham, B. (1980) Grondprijsvorming and
    Grondprijspolitiek, Stenfert Kroese, Deventer
Meijer, H. (1979) Compact Geography of The Netherlands,
    Ministry of Foreign Affairs, The Hague

Ministry of Housing, Physical Planning and Environment (1984) Urban Housing in The Netherlands, Department of Information and International Relations, The Hague

Nationale Woningraad (1985) Woningraad extra, no. 33

Needham, B. (1982) Choosing the Right Policy Instruments, Gower, Aldershot

Needham, B. (1983) 'Local Governments and Industrial Land in England and The Netherlands', in Urban Law and Policy, vol. 6

NIROV (1985) Onderzoek naar de Oorzaken en Achtergronden van de Prijsontwikkeling van Woningkavels, Den Haag

Prins, J.H.A.A. (1985) De Gemeentelijke Grondexploitatie, Vereniging van Nederlandse Gemeenten, Den Haag

Provincie Gelderland (1983) Tussenrapportage over de Financiele Problematiek bij de Exploitatie van Bestemmingsplannen, Provincie Gelderland, Arnhem

Schultz, E. (1983) De Nederlandse Droogmakerijen, Rijksdienst voor de Ijsselmeerpolders, Lelystad

SEO (1979) Onderzoek naar het Gemeentelijk Grondprijsbeleid, Amsterdam

Thomas, D., et al. (1983) Flexibility and Commitment in Planning, Martinus Nijhoff, The Hague

Valk, A. van der (1986) Eindrapport SRO/ZWO Project Planvorming and Uitvoering, Universiteit van Amsterdam, Planologisch en Demografisch Instituut, Amsterdam

Venverloo, A. and Tromp, N. (1972) Gemeentelijk Grondbeleid - Beheer en Administratie, Samson, Alphen aan den Rijn

Zwaag, J. van de (1977) Nederland is Bijna Klaar, Vereniging tot Behoud van het Ijsselmeer, Edam

Chapter Five

FRANCE

Jon Pearsall

INTRODUCTION

In 1945, France was to a large extent still a rural country. It has since become a predominantly urban country, and experienced a 50 per cent increase in population. This change has involved a rapid increase in the housing stock - about half the present housing stock has been built since 1948 - and a large transfer of agricultural land to urban uses. In the 1980s, the rate of population growth has begun to decline, and the emphasis has shifted to improving the design of housing and the quality of the urban environment - in which mistakes were made during the period of rapid growth. There has been a shift from flats to houses, from new development to urban renewal, from tenancy to owner-occupation, and from primarily state-financed to primarily privately-financed housing (Pearsall, 1983).

The whole process of housing development and land acquisition, however, has been - and still is - supervised and to some extent controlled by the French 'State'. The State (l'État) is by no means synonymous with 'Government'. It is rather a national network of public administration, stretching from Paris to the départements and communes (local authorities) and closely linked with a wide range of quasi-public organisations. These organisations, which are neither 'civil service' nor purely 'private', go under a somewhat baffling variety of acronyms (see Glossary), and some of them are specifically concerned with land acquisition and housing development.

In their attitudes to 'the state and the market', the French might at first sight seem paradoxical. France has traditionally had a strong centralised state and, since 1945, it has exercised extensive control over economic activity, including housing development. On the other hand, French men and women adopt a very conservative attitude to landownership, and strongly resist any infringement of their property rights. One Socialist politician has written:

'our enemy is not the aristocracy, but the bourgeoisie: it is not feudal property concepts, but the property concepts of the French Revolution and the Civil Code ... In fact, property is for us a fundamental value and a cardinal virtue; it should be inscribed on the facades of our public buildings with the same emphasis as equality, and well ahead of liberty and fraternity' (Pisani, 1977, p. 10).

This conservative/anarchic streak has sometimes led to successful resistance by landowners to attempts by the authorities, on town planning grounds, to restrict development. At other times, it has led to successful attempts to prevent development, when this was approved by the authorities. It perhaps also explains why the authorities have equipped themselves with extensive powers of pre-emptive land purchase, but prefer to proceed by negotiation and agreement.

Since the 1960s, French policy has become increasingly 'pragmatic'; the authorities have decentralised public administration and given a greater role to the private sector, while not accepting the doctrine that market forces will solve all problems (Ashford, 1982). An extensive range of land policy instruments - compulsory or pre-emptive purchase, development controls, and infra-structure charges - has been developed, in response to the stresses of rapid urbanisation (and, more recently, suburbanisation). After a process of trial and error, these instruments have become an accepted part of the development system. It needs to be stressed, however, that in France there can be a wide gap between written rules and the way things actually work.

In 1945, most urban housing was privately owned, being either owner-occupied or tenanted. The tenanted housing was subject to rent controls - which to some degree have remained. When the State decided on a massive housing programme after the War, it assumed the role of the main provider of housing but it has since steadily withdrawn from this role. The main arm of public sector housing consists of the HLM organisations (Habitations à Loyer Modéré, 'Dwellings at moderate rent'). These were originally philanthropic housing organisations set up in the early 20th century. In 1948, they were reorganised and brought under state regulation, so that state funds could be channelled through them, to provide cheap rented accommodation for lower income groups. They have also built dwellings for sale. The HLM consists today of 1,200 quasi-autonomous organisations, owning 2 m dwellings. The other arm of subsidised housing is the secteur aidé (aided sector) under which private contractors are given assistance for providing cheap rented housing, or housing for sale at subsidised prices.

FRANCE

The construction of HLM dwellings peaked in the 1960s, and the emphasis has since shifted to secteur aidé, and purely private construction, financed by banks and building societies. Private construction now accounts for over half of the total (which peaked at 546,000 units in 1972 and has fallen to under 400,000 p.a. in the 1980s). Even when housing is purely private, however, quasi-public bodies are often involved in land acquisition and preparation, and the existence of rights of pre-emptive purchase influences the process of negotiating land transfer.

FOUR PERIODS

Four periods can be identified in the post-1945 history of urban development in France - although the dates are to some extent arbitrary. In each period, housing policy and land policy have played key roles. The complex relationships between them, however, have at times been clear and consistent, at other times tenuous or inconsistent (Heymann-Doat, 1983).

1945-1953	The Aftermath of War
1953-1963	The Dominant State
1963-1975	Liberalisation
1975-1987	Retrenchment and Streamlining

1945-1953 THE AFTERMATH OF THE WAR

France emerged from the Second World War with a severe housing shortage. War damage had reduced the housing stock, while the influx of displaced persons, compounded by the unprecedented growth of the population and new households, put great pressure on the housing market.

The immediate response by the State was a holding operation. No major housebuilding programme was initiated, apart from the reconstruction programme for the war zones. Anticipated rises in rents were forestalled by the 1948 Rent Act, and public and private investment was directed to rebuilding and modernising the industrial base and infrastructure. Not until after 1953 were major resources made available, directly or indirectly, to increase housing output. And it was not until then that significant changes were made in the prevailing means of ensuring land availability, so as to enable housebuilding to proceed at desired rates.

1953-1963 THE DOMINANT STATE

During and after the 1950s, France experienced historically high levels of population growth, economic growth and urbanisation. Between 1846 and 1946, the urban population had grown from 24 per cent to 50 per cent of the total: from 1954 to 1968, it grew from 56 per cent to 70 per cent, and cities grew at an average rate of 2.7 per cent p.a. (Dyer, 1978). Great pressure was put upon the State to ensure not only that sufficient housing was constructed, but that it was located in the rapidly expanding urban centres - notably around Paris.

The year 1953 was the turning-point in post-war policy, although the ground had already been prepared for the expansion of housing construction. In 1948 the social housing sector, dominated by the HLM organisations, was re-organised, in preparation for its major role in the following decade. State credits on very favourable terms were made available for 'social housing', and cheap mortgage loans were also given, through the Credit Foncier, to encourage private investment in housing for sale.

Land Availability and Expropriation

High construction rates were envisaged, and it was appreciated early on that relying entirely on the market mechanism to release land in adequate quantities, of suitable unit size, and in the desired locations, was unacceptable. In 1953, a new land law extended the applicability of existing rights of expropriation. Since 1810, the State had possessed powers of expropriation for the benefit of the community (d'utilité publique). Under this legislation, land could be expropriated for public infrastructure provision, especially roads and defence works. The 1953 legislation was a major step forward, for it allowed expropriation for housing and industrial development. Land could be expropriated with the intention of selling individual housing plots (lotissements) to developers. Land could also be expropriated for the assembly of large sites which needed to be equipped with basic infrastructure for major housing developments; in the 1950s and 1960s, these developments usually took the form (under the ZUP procedures; see below) of very large estates of high-rise blocks of flats which became famous, or notorious, as the grands ensembles. In addition, land could be acquired for use at a later date for housing or industrial purposes, thus in effect creating land banks, although the term réserve foncière did not become a judicial category until 1967 (Renard, 1980).

In 1958, an ordinance extended the limited definition of what constituted utilité publique to a broader one of intérêt général - general interest. The terminology DUP (declaration d'utilité publique) remains, and so do the procedures of expropriation, appeal and compensation, but far greater power was given to the State, through the Prefect, to determine in what circumstances land and property could be expropriated. It offers great potential for public intervention in land management and urban management in general, although in practice this power has not been widely exercised.

The expropriation procedures, which had been laid down in 1935, were based upon a century of legislation and established practice. They were elaborate and long drawn-out, with rights of appeal, eventually if necessary to the Conseil d'État (constitutional court). To speed up the process of land release, the 1953 legislation also introduced the procedure d'urgence. Under this 'accelerated process', land could be expropriated after the principle of expropriation had been established but before the level of compensation had been determined; the latter commonly caused the most disputes and subsequent delays. Originally, this procedure could only be used to acquire land for social housing and secteur aidé development, but by 1958 all definition of the circumstances in which it could be applied had disappeared, and it was left to the State to judge when it should be used.

The 'accelerated process' proved to be a valuable tool, and during the 1950s nearly all social housing, in the Paris region at least, was built on land acquired under it. In the 1960s and 1970s, the procedure came to be used less and less, and in more restricted circumstances. This was partly because of a reduction in the building programmes, and partly because of objections to the process as such, as the role of the interventionist state became less acceptable.

Administrative arrangements had to be improved for the exercise of the extended powers. Under the 1935 Law, the powers of expropriation were vested in the State, the département or the commune. However, because of the traditional inadequacies of the communes in technical matters, including land acquisition and development, and the reluctance of the central administration to take on the task, powers were made available in 1953 to the OPHLMs (Office Publique d'HLM, the principal arm of the HLMs) and in 1958 to appropriate Sociétés d'Économie Mixtes (SEMs, public/private companies).

The SEMs are legal entities operating under a mix of private and public law, which can be established to carry out specific tasks, including development. They combine some of the prerogatives of the state, including compulsory land acquisition, with the flexibility, skills and finance of the private sector. If a particular company is designated as a concessionaire d'opérations d'aménagement (development agent) it can engage in all aspects of the development process, including land acquisition. Some of the companies are organised on a regional basis and, because of their concentration of expertise and finance, provide an invaluable service to individual communes.

Before 1953, only the State or commune could retain land which had been expropriated. This rule was abandoned, and it was allowed to vest land in an OPHLM, SEM, or even in private bodies building secteur aidé housing. The circumstances in which land could be vested in such bodies were further extended as time went by, and concern about what was happening finally surfaced in the early 1960s.

Compensation

The basis of compensation is critical to any expropriation procedure. Until 1935, a jury of landowners made the final assessment, taking into account a valuation made by the State. This valuation ignored any increase in value consequent on the declaration (or completion) of public works. If land adjacent to the land to be acquired rose markedly in value, that increase would be subtracted from the compensation. Given the make-up of the jury, it was not unexpected that compensation levels were generally in excess of the State's valuation, often twice as much. In 1935 the jury was replaced by a valuation commission, a body markedly less dominated by landed interests. Nevertheless, compensation remained generally higher than the State-determined level.

In 1955, following major rises in land prices as the building programmes got under way, regulations were introduced establishing evaluation formulae and ceilings on compensation. The valuation commission was no longer free to establish compensation as it saw fit. Bitterly opposed, the regulations were legally challenged, and in 1958 the Conseil d'État ruled that they were unconstitutional, undermining the rights of private property enunciated in the 1798 Declaration of the Rights of Man, and incorporated in the Civil Code of 1804. The regulations were withdrawn in 1958, and there has been no serious attempt to reintroduce them (Topalov, 1977).

The 1958 ordinance, at the same time as withdrawing the regulations, replaced the valuation commission with a judge (juge foncier) who, it was hoped, would take a less partisan view of what constituted fair compensation. In practice, judges have generally taken a very generous view of compensation levels. The State is essentially reduced to exhorting them to fix more realistic levels of compensation.

Land-use Plans

The housing construction programme, and the related land acquisition programme, clearly required an overall development strategy, with land-use plans and controls (Heymann-Doat, 1983; Pearsall, 1983). The existing planning legislation (of 1943 and 1945) was conservationist in outlook, and inappropriate for a rapidly urbanising economy. In 1955, nationally applicable development regulations (RNU, Règles Nationals d'Urbanisme) were introduced, which included arrangements for granting development permits. These regulations now underpin all development control, especially where there is no approved local plan.

In 1958, a system of land-use plans was introduced, consisting of Plans d'Urbanisme Directeurs, and Plans d'Urbanisme de Detail (PUDs). The former covered the overall development of towns: the latter more localised areas where detailed proposals were required. At the same time, the ZUPs (Zones à Urbaniser par Priorité, Priority Urbanisation Zones) were introduced. These in fact codified existing arrangements for housing development using expropriation powers. Within these zones, powers of pre-emptive purchase (see below) and expropriation could be, and usually were, exercised, and development was carried out by a SEM or other designated organisation.

The objectives in the setting up of ZUPs - according to the enabling legislation - were to;

(1) create rational development patterns and provide agreeable living conditions, especially for people of moderate income;
(2) coordinate the financing of land acquisition and development;
(3) centralise the provision of infrastructure;
(4) unify the responsibility for development extending over more than one municipality.

The ZUPs became the key to urban development, and developments of more than 100 units outside them were generally refused building permits. In fact, the new planning system (PUDs) was used more as a means of post hoc legitimisation of the ZUPs than as key elements of a coherent system of land-use planning. The legislation of 1958 also introduced the concept of urban redevelopment areas (renovation urbaine) for town centre reconstruction. The financing of ZUP development lay essentially with the State, although a few changes were made in the provisions for pre-emption, expropriation, and housing finance so as to allow some private involvement. It was not until the 1960s, however, that the private sector was actively encouraged to take part.

The financing of land acquisition and infrastructure provision was carried out by two Finance Ministry organisations. The FNAT Fonds d'Aménagement du Territoire (National Planning and Development Fund) provided the finance for land acquisition, and the FDES, Fonds de Developpement Économique et Social (Economic and Social Development Fund) made loans and grants for infrastructure, including schools and public buildings. The availability of funds enabled the development agencies to acquire land at the speed required by the construction process.

Legal constraints upon private development companies effectively excluded them from major construction projects, and two public construction and land preparation companies were established. These are SCET, Société Centrale pur l'Equipement du Territoire (Central Company for Infrastructure Provision) and SCIC, Société Centrale Immobilière de la Caisse des Dépôts (Central Building Company of the Caisse des Depots (a state bank)). Both played a key role in housing construction, and are still active, although less important than in the past. The shareholders of SCET are various public financial agencies, and it has been described as 'a corporation which functions under the aegis of the administration in a spirit of public service' (d'Arcy, 1968). It can provide expertise, and financial contacts, which may not be available locally.

The State took the initiative in housing development at this time, but nevertheless looked to the private sector to contribute to infrastructure costs. Under a decree of 1957 (incorporated in the RNU in 1961), all recipients of a building permit or a permit to develop a lotissement had to make a contribution (participation) to the commune to help cover the costs of infrastructure provision. The levying of these participations was rapidly extended illegally to cover all types of development and, in practice, building permits were, in many instances, 'sold'. In 1967 the Loi d'Orientation Foncière (Land Orientation Law) introduced the TLE (Tax Locale d'Equipement, Local Infrastructure Tax) as a means of bringing order into the situation. The tax can be set at up to 5 per cent of the development cost of the project. Subsequent legislation of 1971 allowed for participations to be negotiated, and the interpretation of this law led to the re-emergence of the practices which the 1967 Law sought to eliminate. The 1985 Loi d'Aménagement) (see below) will, it is intended, bring an end to such abuses (Flockton, 1984).

The 1950s: An Assessment
Almost exclusive emphasis had been placed upon rapid housing construction and a concomitant speedy release of land. new administrative and financial arrangements had been instituted to achieve these ends. Established procedures for land acquisition and release had been significantly modified, although the rights of landowners, with special regard to compensation, remained protected. Housing output had risen dramatically (from 120,000 in 1953 to 298,000 in 1958), and land had been made available to build the housing. The worst of the post-war housing shortage had been dealt with - but there were costs.

Despite the existence of land-use plans, urban development remained largely anarchic. The availability rather than the location of building land was decisive, and many unsuitable places were accepted for building. The ZUPs proved unsatisfactory in terms of scale, design, location and social character. Service provision often arrived years after the dwellings had been completed. In the Paris region, in particular, private developers building for sale avoided the ZUPs, which became one-class developments consisting almost solely of HLM dwellings. The ZUPs had thus not fulfilled all the objectives outlined in the legislation which set them up. At the same time, rent control caused considerable dilapidation in the cities, and urban redevelopment - although not as widespread as in the UK - was marked by insensitivity, especially towards the poorer displaced households.

There were also complaints about rising land prices. In trying to assess the actual rise in land prices from the available data, there are considerable problems. Definitions have changed, and comparisons between different periods can be misleading. On the available statistics, however, the (nominal) price of land in the major towns and cities increased by at least 300 per cent in the 1950s, often 400 per cent, and in the Paris agglomeration by at least 500 per cent. The highest increases were in the period 1950-58; there was a slowing-down towards the end of the decade. These figures hide considerable variations. In Rennes the price of land remained relatively stable, largely because the City and adjacent communes pursued a vigorous policy of land banking; some 85 per cent of all land transactions were undertaken by the authorities.

A measure of the change in land prices in real terms is provided by the charge foncière - the percentage of land costs in the total cost of housing. This generally rose, although not always spectacularly. In Paris, in the years 1957-62, the charge foncière was 18 per cent; in 1963-67, it had risen to 22 per cent. In the suburbs, which were experiencing the greatest growth, the figure rose from 10 per cent to 20 per cent in the same period (Topalov, 1977, 1985; Comby and Renard, 1986).

1963-1975 LIBERALISATION

The year 1958 saw the end of the Fourth Republic and the creation of the Fifth Republic, with De Gaulle as its first President. The Fourth Republic had experienced a continual succession of governments; the change to a presidential system produced greater stability in government, and also ushered in an era - lasting to the present day - in which private enterprise has played a larger role, and in which

administrative power has, to a significant degree, been de-
volved from Paris to the regions and communes.

By 1962, De Gaulle had consolidated his position, and
the next few years saw the relaxation of the existing financial
and administrative regime, and the introduction of more
liberal policies, not only in industry but also in housing,
which was to be opened up to the operation of the market far
more than before. The strategy was to reduce public expenditure (relatively if not absolutely), encourage private investment, and induce tenants and purchasers to make a higher
contribution to housing costs.

This did not mean that the State no longer involved
itself in land and housing policy. Indeed, important inno-
vations were made to improve the availability of land at
reasonable cost, and prevent major speculative development.
In 1962 (followed by consolidating and extending legislation in
1965 and 1971) the Zones d'Aménagement Différé (ZADs,
Deferred Development Zones) were introduced. Within these
zones, the State (and other public bodies) can exercise the
right of pre-emptive purchase of land, at the price prevailing
one year before the sale of the land is proposed. The ZADs
have a life of fourteen years, which is extendable. They were
widely used, particularly in the Paris region but also elsewhere, and especially where major state-inspired investment
and development was planned. The Paris New Towns, the
resort development in Languedoc and the Fos industrial
complex near Marseilles are examples (Strong, 1979). Large
areas were zoned along the planned routes of the A23 and A6
motorways out of Paris, in order to dampen speculation. The
ZADs have also been used for much smaller areas, and for
physically non-contiguous areas in suburban locations, e.g.
in Rennes. The existence of a ZAD has generally dampened
down land price rises, and the system has worked well. The
threat of pre-emption and, more importantly, the vast areas
included and the long life of the schemes, have reduced
speculation, and little actual State acquisition has occurred.
This is in line with the authorities' intentions, for the ZADs
were not regarded as 'land banks'. Moreover, funds were
limited, and were directed into the ZUPs, rather than into
long-term land acquisition.

It was not until 1967 that the Loi d'Orientation Foncière
introduced land banks, or réserves foncières. The State made
funds available to allow communes to purchase land for major
urban developments, tourist developments or leisure areas,
which were not to be established immediately. Financial assistance would be made available to a commune or group of
communes which produced a Programme d'Action Foncière
(PAF, Land Acquisition Programme) - a coherent programme
of land acquisition developed in the context of a development
strategy and land-use plan for the area. The reserves were
seen as complementary to the ZADs, and not as a means of

municipalising land. They were only a means of preventing what could be characterised as excesses of the market (Lenoir, 1977).

The 1967 Law also replaced the ZUP with the Zone d'Aménagement Concerté (ZAC, Zone of Concerted Action). This had the same aims, but was designed to encourage greater private sector involvement in development. Within the zones, pre-emption and expropriation rights can be used to help private developers assemble land. A greater say is also given to the developer in design and layout. The ZACs have been successful in attracting private finance, and display a greater social balance and a better mix of tenure types than most ZUPs.

Land expropriated or pre-empted by the State is eventually disposed of to a private body or individual. Leasehold came late to France, and is not a popular means of holding land. Legislation passed in 1958 allows public bodies to dispose of land on long leases, but has rarely been used. The practice of disposal by (freehold) sale has been criticised from the Right because public powers are being used to transfer land - acquired from one owner at a reduced, non-speculative price - to another owner. It has been criticised from the Left because the State loses both potential future gains and the opportunity to use its landholdings to intervene in the urban land market.

In the 1960s, the Left advocated a policy of municipalisation (Topalov, 1973, 1977). City-wide land offices were to become the sole bodies purchasing land, and were to coordinate purchases with a general urban development strategy. The land would be made available for development on long leases. The Socialist Party is still committed to this policy, but the Mitterand Government made no attempt to implement it.

The rapid growth after 1953 in the number of bodies able to exercise rights of pre-emption or expropriation, and their uncoordinated operation, aroused concern in Government circles; a single central land purchasing agency was proposed. This proposal came to nothing, principally because of widespread local opposition to the establishment of a powerful body run from the centre. A land agency was, however, established for the Paris region, AFTRP (Agence Foncière et Technique de la Region Parisienne); another for the lower Seine; and another for Eastern France. All three are responsible for land acquisition for other public bodies including New Towns; for residential and industrial development; and for the establishment and management of ZADs. The AFTRP alone has the power to engage in development on its own behalf.

The land-use plans (PUDs) of the 1958 legislation were largely ineffective. Even when they were prepared at all, they mainly legitimised the ZUPs and other developments retrospectively. The Loi Foncière of 1967 abolished the PUDs, and introduced SDAU (Schema Directeur d'Aménagement et d'Urbanisme, Structure Plans) and POS (Plan d'Occupation des Sols, Land-use Plans'). The Structure Plans are concerned with general strategy and the Land-use Plans with detailed land use. These plans were to be one of the principal means for coordinating development with public infrastructure investment. The Law was designed to improve the whole development process, by making it market-led, but assisted by the public sector.

Under this legislation, a plot ratio (Coefficient d'Occupation des Sols, COS) is laid down for each area zoned for future development. If this density is exceeded, the developer makes a participation, or monetary contribution, to the commune. The payment must be used for specified purposes around the new development. The COSs are determined locally (being an integral part of a land-use plan) and vary widely. They need to be seen, however, in the context of a nationally-determined density ceiling - Plafond Légal de Densité (PLD), introduced in 1975. The plot-ratio of the PLD was to be 1.5 in Paris, and 1 elsewhere. If the plot ratio of a development exceeds this level, and rises to that permitted under the local plan, the developer has to make a payment - Versement pour le passement du Plafond Légal de Densité (VPLD) - to the commune. The money must be used for specific development purposes in the commune.

Land Taxation

The aim of VPLD is two-fold - to help the communes improve the general standards of development and facilities in their areas, and to act as a tax on 'speculation' and excessive profits in areas of greatest development pressure. Like some other administrative and fiscal arrangements, the PLD arrangements have rapidly become out-dated. In the pre-1975 period of rapid economic growth, a housing boom, and a healthy construction industry, they may have been appropriate; in present circumstances they are not, and have recently been relaxed (see below). As the VLPD is in effect a land tax, it is convenient at this point to discuss land registration and taxation. All land-holdings are registered with the Ministry of Finance, at its departmental offices, on a plan cadastral, which indicates ownership, area and land-use. This 'Land Register' is used as a basis for land taxes, but not for legal proof of ownership. The oldest land taxes are annual taxes on unbuilt land (impôt foncière sur les propietés non bâties) and on built land (impôt foncière sur les propietés baties). They were designed in the 19th century and bear no relation to the 20th-century situation. Agricultural land is undervalued relative to built-up land, and the situation is compounded by the fact that revaluations are infrequent, so that valuations are often wildly out of line with market values. It is not uncommon for land zoned for future building to be taxed on an agricultural basis, and for genuine agricultural land to be more heavily taxed. The tax was designed as a source of revenue for local authorities, but in practice it is minor (10-15 per cent of receipts). It has had undesirable effects on land release and the urbanisation pattern (Flockton, 1984).

The land tax system needs to be reformed so as to be more consistent with land-use zoning, to recoup some betterment, stimulate land release, and provide more revenue. Indeed, the introduction of the taxe locale d'équipement was partly a response to the feebleness of the old land taxes. In 1966, the Rapport Bardier (Bardier, 1966) recommended a single annual land tax for all land, based upon market value. In 1967, a more limited version was incorporated in the Loi d'Orientation Foncière in the form of a taxe d'urbanisation. This was an increased tax on land zoned for building within an approved POS. It was seen at the time as an important element in the battery of land instruments, but it was defeated by political and bureaucratic pressures, and was withdrawn. More limited measures were considered in 1974, 1975, 1977 and 1978, also with no results (Flockton, 1984).

During this period (1963 to 1975) there was a change in the behaviour of land prices. There was a marked slackening-off in the general rate of increase, partly perhaps because of the impact of some of the public land-use measures, but also because of a slowing of demographic growth and urbanisation. There were, however, marked spatial differences, especially between central city districts and suburbs. In many central city districts, there was a stabilisation of prices, resulting from the impact of the PLD and a decline in population, while in some suburbs, especially those of a more rural type, prices continued to rise.

RETRENCHMENT AND STREAMLINING 1975-1986

From the mid-1970s onwards, the world economic crisis was accompanied by a decline in the rate of population growth and the emergence of new patterns of urbanisation. Annual housing construction peaked in 1974, and fell from then on. By the end of the 1970s, there was a talk of a housing crisis, insufficient housebuilding, rising credit costs, and an insufficiency of land available for building - especially in the Paris region. The concern of the 1950s and 1960s with the problems of rapid growth, and the consequent administrative, financial and fiscal responses, were increasingly perceived as being irrelevant. As usual, the administrative response to these changes lagged behind events, and a number of measures introduced in the late 1970s were clearly rooted in the problems of the 1960s and early 1970s.

The Loi Foncière of 1975 - among other measures - introduced the ZIF (Zone d'Intervention Foncière, Zone of Land-use Intervention) in which authorities were able to pre-empt the purchase of any urban land, although (unlike the provisions in the ZADs) only at market prices. The procedure was introduced to enable the authorities to improve and restructure existing urban areas, especially town centres. In 1975, the Nora Report recommended that greater emphasis should be placed on the rehabilitation of dwellings, especially in town centres, rather than on redevelopment or new building (Ministère du Finance, 1975). This change of emphasis was for economic and social reasons; scattered, 'peri-urban' development was being matched by social segregation in town centres.

In 1976, the Fonds d'Aménagement Urbain (Urban Management Fund) was established, with the task of encouraging and partially funding approved schemes of city centre rehabilitation and management. The introduction of the VPLD was seen as part of this strategy. The existing roles of the PAF (Programme d'Action Foncière, Land Acquisition Programme) and, where appropriate, the AFU, were maintained. The AFUs (Associations Foncières Urbaines, Urban Land Associations) are local associations sponsored by the State, which bring together land and property owners in order to pool land and generally facilitate agreement on development and redevelopment (Duban, 1982).

Legislation in 1976 introduced Zones d'Énvironnement Protégé (ZEP, Environmental Protection Zones) which could be established in rural areas to protect landscapes and buildings of special architectural or aesthetic value. In addition, the legislation on Perimetres Sensibles (Sensitive Zones), introduced in 1959 to allow public bodies to exercise pre-emption powers in order to ensure public access to woodlands etc., was strengthened. In 1977 a circular recommended that, where a POS was not in effect then, in interpreting the 'national ground rules' (RNU), there should be a pre-disposition against development unless it could be shown to be appropriate.

The rate of preparation and approval of local plans (POS) proved to be slower than anticipated. Nevertheless increasing numbers were approved, particularly in urban and 'peri-urban' (fringe) areas where there was most pressure for housing land. On average, only one-sixth of land within a POS is zoned for development of any sort. The POS, by their very nature, were thus more restrictive than the previous situation of inadequate PUDs and reliance on local interpretation of the 'national ground rules' (RNU).

In the late 1970s, the political climate regarding urban development was contradictory. There was a reaction against the gigantism of the 1960s, and a fashionable concern with the environment. Restrictive policies were promulgated at national level, but also became very common at local level. People dispersing from city centres and older suburbs to the periphery and outlying villages took political control of the communities and resisted any further development of 'their' village. This resistance may or may not be supported by official estimates of land requirements and land availability, and these estimates are naturally used or challenged by developers and residents alike, according to whether the estimates accord with their own inclinations.

But at the same time as the authorities were adopting restrictive policies, they were encouraging housing development which made increased demands on land. The latent demand for maisons individuelles (two-storey detached or semi-detached dwellings), as opposed to collectif development (blocks of flats) was being met by private developers. Much of this development was financed privately, but the 1977 reform of housing finance gave specific incentives to the secteur aidé to build houses rather than flats. The Mayoux Report commented on the problems posed by such development on the urban periphery, but accepted the inevitability of it (Ministère de l'Énvironnement, 1979).

In the liberal philosophy of the Giscard Government, the role of the State was seen as neutral. The State should intervene only in establishing land reserves or pre-empting land sales in strategic situations. Making land available for development should be done privately; the sort of intervention undertaken in the 1950s and 1960s was considered no longer appropriate.

The Saglio Report (Ministère de l'Énvironnement, 1980) came to similar conclusions. Specifically requested to examine the question of land availability, the authors indicated that the state should improve the operation of the existing land acquisition and planning mechanisms, and act only as an intermediary in the operation of the land market. In the short term, spare land in existing ZACs, ZUPs and New Towns should be brought forward for development, while the processing of lotissement documents should be speeded up. In addition, land held by public bodies such as HLMs, Caisse des Dépôts and, in the Paris region, the AFTRP, should be released. In the longer term, land committees should be established in the Departments, consisting of all private and public agents concerned with the local land market; their function would be to gather information and publicise it. Moreover, land organisations like the AFRDP should be established in a dozen metropolitan areas. The procedures for considering development proposals needed to be made more efficient, and the operation of local plans (POS) needed to be made more responsive to changing needs. Finally, owners should be encouraged to release land; special attention needed to be paid to a reform of the land tax on undeveloped land.

The Socialist Government

In 1981 the Socialists won a sweeping Presidential and Parliamentary victory. For a time, the Government sought to spend its way out of the recession; public expenditure for the construction industry, including housing, was increased, and interest rates reduced. Unable on its own to sustain this policy, France was forced to revert to a liberal economic policy, including cuts in public expenditure.

There were no radical departures in housing policy, the 1977 reforms remaining largely unchanged. The APL (Aide Personalisée au Logement, Personal Housing Allowance), introduced to help lower income householders to rent or purchase dwellings, were kept up with inflation, although the charges for heating, lighting and other services, which lie outside it, became increasingly burdensome. Interest rates on state loans were trimmed but remained close to market rates. In the uncontrolled sector, or secteur libre, interest rates remained high. Rents remained controlled, even if loosely, after the de facto collapse (certainly outside the high-value sector) of the Quillot scheme of 1982 for bringing landlords and tenants together to decide rent levels and contractual obligations by mutual agreement. Marginal changes were made to the availability of subsidised loans (they can now be used to purchase second homes), and the process of approving loans is supposed to have been streamlined and speeded up. A pragmatic improvement and simplification in the operation of the existing legislation, administrative procedures, and finance provisions, was the main characteristic of the Socialist regime. Its attitude to 'the market' was somewhat ambivalent; it began by extending rent control, but it had no wish to kill off private renting, and subsequently relaxed the legislation concerning rents.

In 1983, the Loi de Finances relaxed the PLD (plot ratio) control and allowed communes with POS to double the plot ratio for their area if they so wished. All public buildings are now excluded from its operation. The plot-ratio (COS) can also be relaxed if the communes require it. Most of the highly urbanised communes have taken advantage of the change; the results should be more high density housing; more extensions to, and improvements of, the existing stock; and a boost to the construction industry. Land will be released and will rise in price in the most favoured areas such as Paris and Rhône-Alpes/Provence, and some pressure will be taken off 'peri-urban' development.

The Socialist Government also attempted to tackle the problems of the 'old' land tax, by introducing up-dated valuations. Action was formally started in 1982, but technical and unstated political difficulties made the process very slow. There is no reason to believe - especially since the election of the conservative Chirac Government - that there will be an improvement on previous administrations' performance on this matter.

The principal legislation of the Socialist regime which is directly concerned with land issues was the Loi d'Aménagement of 1985. The law did not introduce any radical new principles, but built upon the established approach of simplifying procedures and 'reducing bureaucracy', so as to get better planning and involve the citizens more closely. This strategy was seen as necessary because the increased complexity of procedures had slowed up the planning and development process unduly. The legislation also increased the powers and duties of the communes in the field of land-use planning, which reflects the decentralisation process going on in France.

The Chirac Government

In the 1986 elections the conservative parties won a majority in the Chamber of Deputies, so that a conservative Prime Minister, M. Chirac, was elected - although M. Mitterand remained President. The Chirac Government is philosophically attached to deregulation (and privatisation) and has had no difficulties in taking up the principles, and indeed the substance, of the Loi d'Aménagement. The principal tool of land policy remains the right of pre-emptive purchase, as available in the ZADs and ZACs, but a number of changes have been made. A new 'Urban pre-emption right' (DPU, doit de pre-émption urbain) has been introduced, replacing the ZIF procedures. The DPU procedures are not unlike those of the ZIF but they have been streamlined, which should result in speedier and more satisfactory results. Compulsory acquisition powers remain as before. The scope of the ZAD has been slightly restricted; it can be used only in rural areas not subject to a POS, and then only at the discretion of the Prefect.

The new law also seeks to tackle the problem of the abuses of the participations made by developers. Instead of ad hoc arrangements between communes and developers (in effect, the sale of building permits), the commune can establish a perimètre-fixé (fixed perimeter). This is a line drawn around a specified area, within which the rights and obligations of developers are defined, and made public. Developers will thus know their obligations in advance; if more is demanded of them by the commune, they have the right of appeal to an independent tribunal, without fear of being refused a permit. Outside these defined areas, permits can no longer be made conditional on financial contributions. The TLE (infrastructure tax) arrangements remain in force. For comprehensive development, the ZAC remains the principal tool; indeed, its use has been extended, replacing the cumbersome reno- vation urbaine procedures. Certain changes have been made to allow the associations of property-owners (AFUs) to function more freely.

LAND PRICES

Land prices, in money terms, have increased steadily since 1975. Between 1978 and 1981, for instance, they rose by 20 per cent per annum, whereas construction costs rose by less than 11 per cent per annum. But average land prices conceal considerable variation, the largest increases being in medium-sized towns such as Brest, Dijon and Tours rather than in the major agglomerations. The largest increases are found in the peri-urban or semi-rural areas beyond the suburbs estab-lished in the 1950s and 1960s. However, the land cost as a percentage of the total cost of new dwellings (charge foncière) has remained remarkably stable from the 1960s through into the 1980s. Table 5.1, although restricted to the Ile-de-France and to certain types of dwelling, gives a typical picture of the various cost elements in new housing. The apparent contradiction between the static land cost percentage and increasing land prices is largely explained by the fact that most development is taking place on peripheral areas where land prices are lower.

Table 5.1: Land Costs as a Percentage of New Apartment
Costs (charge foncière). Ile-de-France, 1975-83

Costs		Non aidé			Aidé	
	1975	1980	1983	1975	1980	1983
		%			%	
Charge foncière	18.9	21.1	22.2	17.7	15.0	16.6
Construction	43.7	41.7	45.6	52.2	54.0	53.9
Marketing	30.6	31.0	28.6	24.4	25.1	24.8
Profit	6.8	6.2	3.6	5.7	5.9	4.7
SALE PRICE	100.0	100.0	100.0	100.0	100.0	100.0

Source: Comby and Renard, 1986

Notes: Non Aidé – private sector finance, higher income
                     groups, lower densities
          Aidé   – state-assisted finance, middle/lower income
                     groups, higher densities

CONCLUSIONS

Over the last forty years, a variety of procedures have been made available to the public and para-public authorities (both national and local) for the acquisition of land, and the control and direction of land use. These procedures, both financial and administrative, have been developed incrementally in response to prevailing conditions, and have tended to remain in place even when conditions change. Modifications and adjustments have, however, been made, and there has on the whole been a remarkable continuity of policy, in spite of changes of government. The sort of vexatious and divisive disputes over betterment which have wracked British land policy over the same period find no equivalent in France.

In the 1950s, the emphasis was on large-scale development by quasi-public bodies. From the 1960s onwards, private developers were given a much larger role, but nearly all development of more than a few houses still involved the collaboration of public bodies. This principle has remained unchallenged, in spite of detailed changes. The rights of private property enshrined in the constitution remain staunchly defended, and generally respected. Pre-emption rights are essentially designed to give the authorities a strengthened position in the land market, not as a tool to replace the market. When land is acquired compulsorily, compensation is usually at more generous levels than the State would have considered reasonable. Land-use control, exercised through the system of construction permits issued in the context of land-use plans, has been largely accepted, and so also has the principle of density control. On the other hand, the situation concerning the annual property taxes is anomalous and unsatisfactory, and yet reform has been frustrated by opposition from property-owners.

In drawing conclusions as to the effectiveness of the various procedures, especially in relation to housing, any generalisations are bound to conceal great variations in what has happened at different times and in different places. Distinctions have to be made between the 1950s and the 1980s; between major cities and smaller towns; between city centres and suburbs; and between the various regions of France. Further distinctions have to be made between types of dwellings and sources of finance. A basic distinction is often made between the Ile-de-France (Paris region) and the rest of the country, but this broad distinction hides a varied and changing pattern of demographic and economic growth and decline (Dyer, 1978). There are declining areas - such as the Massif central and the old industrial towns of the North-east - which are characterised by low property prices and some dereliction; there are growing areas - no longer solely in the Paris region - where the problems are of shortage and high prices.

Generalisations also hide the differing responses of the communes. The national legislation lays down procedures, powers and financial incentives intended to put in place a coherent land policy and land-use control system throughout France. It takes the form, however, of powers which the communes may use, if they wish. The larger communes, in particular, have a considerable degree of autonomy, and have made use of the available powers with varying degrees of enthusiasm. It is twenty years since the land-use plan system (POS) was instituted, and still not all of France is covered by it. Some communes have made vigorous use of land banking powers, and have drawn on state assistance to the maximum, for the implementation of their local land policy. Others have not used their powers at all. Overall, the quality and effectiveness of land-use planning has improved in France over the last two decades. Nevertheless, the full potential of the powers available has not been realised. The huge variations in political standpoints and technical and professional capacity among the 35,000 communes makes this inevitable.

As regards the quality of housing development, the position is mixed. The housing in the ZACs is generally considered to be superior to that in the now defunct ZUPs, because the ZACs are smaller in scale, better designed and managed, and usually contain a good mix of tenure and price. The majority of new dwellings, however, are built outside ZACs, either as speculative housing estates, lotissements, or as single units on scattered plots. The location, if not the design, of these developments often leaves a lot to be desired. The frequent failure to exercise effective planning control over such developments results in major infrastructure costs for the communes (ADEF, 1986; Pagès, 1980).

Although the French are better housed than they were in the 1950s, new problems have arisen. French cities have for the most part avoided 'central city' problems of the scale of the USA, but social segregation is becoming more marked. Gentrification of the more attractive town and city centres is widespread, and those authorities who have tried to restrain this process have been held back by the cost of exercising the available powers. Peripheral development of cheaper dwellings has encouraged home-ownership, though with increased costs to the occupiers measured in travel time and transport costs to work. Under present economic circumstances, increasing numbers of house purchasers are finding the move to owner-occupation over-burdensome.

As to the question, 'Is there a shortage of land for housing?', there can be no definite answer. The official view, expressed in the Saglio Report, remains largely unchanged (Ministère de l'Énvironnement, 1980). This is that the existing system is basically sound, and only needs to be made a little more flexible to ensure that adequate land is made available. The private developers, on the other hand, argue that there is insufficient land available. The fact that the charge foncière has broadly remained stable over the last twenty years perhaps indicates that a reasonable balance has been kept between public policy and market demand. The existing problems of access to housing for disadvantaged social groups are less a problem of land policy than of a system of housing finance and management which tends to favour certain groups rather than others. The 'social housing' organisations tend to class as 'bad tenants' the very poor, the handicapped, single-parent families and, above all, immigrants.

The current array of land policy instruments has been developed and exercised in the context of the French cultural heritage. In spite of the dangers of generalisation, one can say that it has been, and continues to be, operated for the most part in a balanced and effective way.

REFERENCES

Ashford, D. (1982) British Dogmatism and French Pragmatism, London

Association pour des Études Foncières (ADEF) (1986) Produire des Terrains à bâtir, Paris: ADEF

Bardier Report (1966), Commissariat General du Plan, La Rapport du Groupe VI de la Commission de l'Equipment urbain du V Plan, Paris: La Documentation Française

Comby, J. and Renard, V. (1986) 40 ans de Politique Foncière, Paris: Éditions Economica

D'Arcy, F. (1968) Structures Administratives et Urbanisation, Paris: Berger Levrault

Duban, Bernard (1982) Les Promoteurs Constructeurs, Paris: "Que sais-je" Presses Universitaires de France

Dyer, C. (1978) Population and Society in Twentieth Century France, London: Hodder and Stoughton

Flockton, C. (1983) 'Local Government Reform and Urban Planning in France', Local Government Studies, 19.5

Flockton, C. (1984) An Assessment of French Urban Land Policy, Department of Linguistics and International Studies, University of Surrey. Typewritten

Heymann-Doat, A. (1983) 'L'Evolution du Droit de L'Urbanisme en France', in Heymann-Doat (ed.) Politiques Urbaines Comparées. Paris: A L'Enseigne de L'Arbre Verdoyant

Lenoir, N. (1977) Les Réserves Foncières, Notes et Etudes Documentaires No. 4375. Paris: La Documentation Française

Ministère de l'Énvironnement et du Cadre de Vie (1979) Demain, L'Espace: L'Habitat Individuel peri-urbain (Mayoux Report). Paris: La Documentation Francaise

Ministère de l'Énvironnement et du Cadre de Vie (1980).

L'Offre Foncière (Saglio Report). Paris: La Documentation Française

Ministère du Finance et Ministère de l'Equipement (1975).
L'Amélioration de l'Habitat Ancien (Nora Report). Paris:
La Documentation Française

Pagès, M. (1980) La Maîtrise de la Croissance Urbaine. Paris:
"Que sais-je" Presses Universitaires de France

Pearsall, Jon (1983) 'France', in M. Wyn (ed.) Housing in
Europe, London: Croom Helm

Pisani, E. (1977) Utopie Foncière, Paris, Gallimard

Renard, V. (1980) Plans d'Urbanisme et Justice Foncière.
Paris: Presses Universitaires de France

Strong, A. (1979). Land Banking: European Reality.
American Prospect. Baltimore and London: The Johns
Hopkins University Press

Topalov, C. (1973) Les Promoteurs Immobiliers: Contribution
à Analyse de la Production Capitaliste du Logement en
France, Paris: Mouton

Topalov, C. (1977) Expropriation and Pre-emption Publique.
Paris: Editions de CRU

Topalov, C. (1985) 'Prices, Profits and Rents in Residential
Development: France 1960-1980', in M. Ball, V.
Bentivenga, M. Edwards, M. Folin (eds.), Land Rent,
Housing and Urban Planning: A European Perspective.
London: Croom Helm

GLOSSARY

AFRTP Agence Foncière et Technique de la Region Parisienne
(Paris region land agency)

AFU Association Foncière Urbaine Urban Land Association (a
cooperative grouping of landowners)

APL Aide Personalisée au Logement Personal Housing Allowance

COS Coefficient d'Occupation des Sols Plot ratio (ratio of
floor space of building to area of site)

CFF Credit Foncier de France (State bank for funding
public works)

DUP Declaration d'Utilité Publique 'for the public interest'
(Basis for expropriation)

DPU Droit de pré-emption urbain Urban land pre-emption
right

FAU Fonds d'Aménagement Urbain Urban management fund

FDES Fonds de Développement Economique et Social Economic
and Social Development Fund

FNAT Fonds d'Aménagement du Territoire Land Management
Fund (a public fund available to communes)

HLM Habitations à Loyer Moderé Dwellings at moderate rent
(social housing)

FRANCE

PAF	Programme d'Action Foncière Land Acquisition Programme
PLD	Plafond Legal d'Occupation des Sols Ceiling density (National maximum plot ratio)
POS	Plan d'Occupation des Sols Local land-use plan
PUD	Plan d'Urbanisme Directeur Structure Plan (replaced by SDAU). Plan d'Urbanisme de Détail Land-use plan (replaced by POS)
RNU	Règles Nationals d'Urbanisme National rules for development control
SEM	Societé d'Economie Mixte Mixed company (public-private development company)
SDAU	Schema Directeur d'Aménagement et d'Urbanisme Structure Plan
SCET	Société Centrale d'Équipement du Territoire (central company for infrastructure provision)
SCIC	Société Civile Immobilière de la Caisse des Dépôts (State development company)
TLE	Taxe Locale d'Équipement Local infrastructure tax
VPLD	Versement pour le passement du Plafond Legal de Densité Payment for exceeding the national ceiling density
ZAC	Zone d'Aménagement Concerté Zone of concerted action (Area in which expropriation and pre-emption rights can be used to assist private developers)
ZAD	Zone d'Aménagement Differé Deferred Development Zone (Large areas in which public bodies can exercise special pre-emption rights)
ZEP	Zone d'Énvironnement Protegée Environmental Protection Zone
ZIF	Zone d'Intervention Foncière Land-use Intervention Zone (urban area where public bodies can exercise pre-emption at market price)
ZUP	Zone à Urbanisé par Priorité Priority urbanisation zone (replaced by ZAC)

Chapter Six

YUGOSLAVIA

Georgia Grzan-Butina

INTRODUCTION

At the present time, there are two systems of land development in Yugoslavia; the formal sector, which is deeply rooted
in the current structure of town planning practice; and the
informal sector, related to the private land market and the
provision of self-build housing. The formal sector, which
comprises the greater part of land development for housing,
operates at two levels: city and regional. There are at present no national policies regarding land development for
housing; exceptions are disaster settlements, the provision of
major national facilities (power plants, airports, highways,
etc.) and development for tourism purposes. In contrast to
many countries, Yugoslavia has no shortage of land; the main
problem is the proper use of land, especially for housing
development.

BACKGROUND

As a modern state, Yugoslavia has a very short history. It
was founded as the 'Kingdom of Serbs, Croats and Slovenes'
(subsequently the Kingdom of Yugoslavia) in 1918. The
present Federal Socialist Republic of Yugoslavia (Federativna
Socijalisticka Republika Jugoslavia) was proclaimed in 1945
after the liberation from German troops and, ever since then,
the country has maintained its non-aligned independent
status.
    The present Yugoslavia has an area of 98,766 square
miles, i.e. approximately the size of the UK, but with a much
smaller population of around 18m. It lies between the Adriatic
Sea and seven inland neighbours: Italy, Austria, Hungary,
Greece, Romania, Bulgaria and Albania. Yugoslavia forms the
north-western part of the Balkan Peninsula, and almost
two-thirds of its territory is mountainous. The most fertile
flatlands, containing alluvial soils, lie north of the Sava

River, towards the Hungarian and Romanian borders. Those areas which do not have favourable agricultural environments have mineral resources such as iron, copper, lead, chromite and many others.

The South Slavs had no common history until their unification. There are today six autonomous republics of the Federal Socialist Republic of Yugoslavia. These are Serbia (including the autonomous regions of Vojvodina and Kosovo Metohija), Croatia (including the ancient province of Dalmatia), Slovenia, Bosnia-Hercegovia, Montenegro and Macedonia.

Post-war Years

After the unification in 1918, the main problem of the newly formed state was to find a form of government which would be acceptable to many national groups. Complete stability was not achieved during the inter-War period, and there were frequent changes in the country's constitution. At the outbreak of the Second World War in 1939, Yugoslavia was formally neutral, although under strong pressure to support Germany; it was also experiencing internal political crisis. On 6 April 1941, Germany invaded Yugoslavia and, with the King and other political leaders already in exile, the Yugoslav army capitulated on 17 April. The country was subsequently divided between various Fascist groups. Between 1941 and 1945, the Yugoslav partisan groups, which were formed by the Communist Party under the leadership of the late President Tito, fought against the German army and local Fascist groups. After the liberation in 1945, Yugoslavia was politically organised into the present socialist federation.

From 1945 until 1956, the whole country was administered by strong central government bodies, but their functions were gradually decentralised to regions, cities and communes. After 1956 most social, cultural, administrative and working units were re-organised into the Basic Organisations of Associated Labour (OOUR, Osnovne Organizacije Udruzenog Rada). However, private businesses are permitted, if they do not employ more than five persons. This decentralised system, based upon workers' cooperatives, residents' groups and commune administrations, finally affected the overall urban and town planning system; self-management and public participation became essential elements in local and regional organisation (Grzan-Butina, 1982).

HISTORICAL FACTORS IN LAND DEVELOPMENT

Land development in Yugoslavia has always been subject to local control. Although there is evidence of Roman and early Slav developments (6th-10th centuries AD) throughout

Yugoslavia, it was not until the Middle Ages that tight control over land development was introduced by many Town Councils (Grzan-Butina, 1983). This control coincided with the establishment of medieval towns, and the introduction of charters guaranteeing towns the right to trade and to govern their own affairs.

Deeds of Property Transfer Records (Protocollum Fassionum) and similar other legal records (court rolls, Statut Publica, etc.) indicate that there was strong administrative control over land development within towns. Control was exercised over the location of particular types of property, the amount of developable land per resident, and the general condition of properties. Major cities levied a local property tax and, from the 15th century onwards, many cities introduced Land Registers. Recent investigations of some medieval towns have shown that a tight control was especially evident in central areas. These central areas generally showed regular, quite often gridiron-type, development. At the same time, other peripheral parts of towns followed a more spontaneous, organic-type development. The size of individual plots varied, depending upon the location within a town and the social status of each resident. A strong residential and economic hierarchy was evident in the variety of building types and their location within a town.

During the 17th and 18th centuries, most Yugoslav towns experienced economic, social and administrative changes, which were reflected in the system of land development. Town Councils started to act as major promoters and controllers of development, and wealthier burghers, merchants and nobility (usually the Town Councillors themselves) were given a right to occupy central locations. They were also allowed to acquire larger plots of land (through the purchase of multiple plots) and thus build larger houses. This process of urban transformation subsequently pushed less wealthy people towards the peripheral parts of towns, and finally brought about the creation of the first suburban developments.

Although internal urban affairs were controlled locally, all major political and economic decisions were initiated by much stronger foreign governments (of either Austria or Hungary), leaving the country impoverished and underdeveloped. These foreign governments exercised their power in Slovenia, Croatia, Bosnia and Hercegovina, which were eventually absorbed by the Austro-Hungarian Empire. Serbia, with Montenegro, was able to retain its political independence, but remained equally undeveloped due to many internal disputes and a lack of proper government (Auty, 1965).

19th-century Urbanisation
During the 19th century, increased urbanisation introduced a number of changes into the economic, administrative and

social structure of most regions and cities. The development
of large tenement buildings, the most popular building type of
that period, was initiated either by City Councils or by
individual private developers. Commercial, office and residential uses were usually combined within tenement blocks.

In order to 'regulate' this rapid urban growth, a number
of master plans were prepared for larger cities such as
Zagreb, Ljubljana and Belgrade (Peric, 1985). The 1864 plan
for Zagreb was based upon purely technical and engineering
principles; plans were laid down for the road layout and
infrastructure, but there were no detailed plans for the
layout of housing. There were, however, detailed plans for
public buildings. Some other plans such as the 1905 Master
Plan for the city of Ljubljana, took into consideration a
number of aesthetic principles in order to enhance the
character of particular historic areas.

After the unification of Serbs, Croats and Slovenes into
the Kingdom of Yugoslavia in 1918, most larger cities continued to control land development in the same manner as
before. Development was regulated by master plans involving
the commissioning of Yugoslav-born rather than foreign
architects and engineers. Private developers, however, still
built residential tenement blocks in the same manner as during the last quarter of the 19th century. City Councils, on
the other hand, were mainly involved in the development
process for the provision of public buildings and the creation
of amenity and open public spaces. Several major cities, such
as Zagreb and Ljubjlana, prepared master plans along the
lines of contemporary town planning ideas; due to the lack of
financial resources, these plans were not implemented until
after the Second World War.

The Post-1945 Period
Land development in post-War Yugoslavia has been closely
linked to the economic, political and administrative development of the country. In order to utilise and build upon the
existing infrastructure, the initial post-war national policies
for development were oriented towards expanding major cities.
This subsequently brought about a greater concentration of
employment, and attracted the working population into the
cities of Belgrade and Zagreb. On the other hand, Ljubljana,
the capital city of Slovenia, has developed differently; it has
been protected from this immigration and has specialised in
higher education, tertiary industry and research activities.

After 1956, when the whole country underwent a
decentralisation of all major industrial, administrative and
other functions, individual cities and regions developed their
own housing policies and programmes. These were introduced
in order to speed up housing provision, and to deal more
adequately with specific problems. The national goal was to

provide housing for everybody as quickly as possible. Zagreb, for example, was growing at the rate of 10,000 residents a year (having 650,000 residents in 1976) and the housing provision had to meet this demand. In addition, new houses were needed to replace damaged or deteriorated older housing.

In order to deal adequately with the emerging urban problems, a new town planning structure was gradually developed. It embodies the decentralised system of the country, and the involvement of local residents at all levels of planning and urban development.

THE PLANNING FRAMEWORK, HOUSING AND URBAN POLICIES FOR LAND DEVELOPMENT

Planning and Urbanisation

There is a very strong relationship between the local administrative system, land development and the system of urban planning and design. In contrast to the British planning structure, which instigates most significant urban policies at the governmental level, channelling them downwards into the local authorities, the Yugoslav system operates the other way around, being most significant at the Commune level. In order to understand the manner in which land is developed, it is important to describe the current structure of planning and urban development.

At present, there are three levels of urban planning: national, regional and city. National planning for the development of the whole country contains basic guidelines and policies that each city or region has to follow. It also includes the setting of income tax levels. The tax system is different from the British one in that tax rates are set for specific sectors, such as defence, intrastructure, health, housing etc. Discussions and referenda are held on the level of tax for these sectors; in some cases, the levels can be varied from region to region.

The Commune (Opcina, Obstina, Komuna) is the basic political and administrative unit; it usually corresponds with a historically developed settlement. Communes can vary considerably in size, although they are usually of 35,000-70,000 people. In the very large cities, there are several communes (Zagreb has ten) and there is a two-way relationship between the city administration and the communes. The communes can make proposals - e.g. on roads or the pattern of development - and they can also block proposals from the city administration. Decisions often involve a long bargaining process.

Urban planning takes place at (a) the regional level and (b) the city level. These two levels have - in theory - been integrated since 1971. Various organisations are responsible for the preparation and implementation of regional and city plans. There are two main types of organisation: Instituti (institutes) which are more policy oriented, and Zavodi (city and local planning units) which are more concerned with detailed land-use and design planning. The preparation of plans is put out to tender to these organisations by the various levels of government. The regular funding of the Institutes and Zavodi is limited to about half their income; the rest has to be earned from fees, in competition with other organisations.

The real urban planning and design starts at the operational level and involves the preparation of various documents and plans:

(i)   The Regional Plan (Regionalni Plan) lays down development policies for the whole region.

(ii)   The General Urban Programme (Generalni Urbanisticki Program) defines broader social, political, economic and developmental objectives which serve as guidelines for urban planning in a particular commune or the whole city.

(iii)   The General Urban Plan (Generalni Urbanisticki Plan, GUP) is the official planning document, which consists of land-use and development plans, statistical information, developmental objectives etc. (approximately equivalent to the British Structure Plan).

(iv)   The Detailed Urban Programme (Detaljni Urbanisticki Program) is similar to 'planning and urban design briefs'.

(v)   The Detailed Urban Plan (Detaljni Urbanisticki Plan, DUP) is a three-dimensional plan indicating the development of areas at urban design scale (approximately equivalent to Local or District Plans).

(vi)   The Implementation Plan (Provedbeni Urbanisticki Plan, PUP). Implementation starts when mutual agreement is reached at all levels and among all working organisations, residents and the commune.

Under the New Constitution (1974), the preparation of the General Urban Programme is the most important aspect of the operational level of planning. The General Urban Programme consists of proposals for housing, commerce, leisure etc. These proposals can be specially prepared by professional teams at city or regional level, or initiated by the communes themselves. When the majority of city residents and the Executive City Assembly accepts the Programme, it becomes the basis for the preparation of the General Urban Plan (GUP).

Every city or town has the autonomous power to develop its own GUP, which has, however, to be in accordance with general economic, social and political developmental policies laid down at National and Commune levels. The preparation of the GUP starts with the public announcement by the City Council (on television, in newspapers and other media) that one of the city Institutes or <u>Zavodi</u> is to be officially responsible for its execution. A special team consisting of various urban professionals (town planners, architects, economists, urban designers, social researchers, etc.) is appointed by the institute to carry out the project.

The final report outlines various programmes, and also includes special reports dealing with particular aspects such as conservation, townscape analysis, historical surveys etc. The GUP normally puts forward alternative options for future development, and serves as a basic guideline for the preparation of Detail Urban Plans, for special architectural and urban design projects, for the location of particular uses, and for the issue of building permits. It takes approximately three years to complete the final document.

The time span covered by a GUP varies from one city to another. In Zagreb, the GUP was initially prepared for a period of ten years (1970-80); in Ljubljana for thirty-five years (1965-2000). Plans which cover long periods, like that for Ljubljana, have proved to be inadequate for proper developmental guidance, notwithstanding the provision for interim reviews. According to the current law, GUPs should be revised and expanded every five years, but because of financial constraints, they are more likely to be revised every ten years. Once completed, the GUP is available for public discussion; the following diagram illustrates the phases and timing generally required for this discussion process and for the final approval of the document.

THE PREPARATION AND EXECUTION OF THE
GENERAL URBAN PLAN

1.	Each Commune produces a number of master plans and proposals.	
2.	All master plans are put together for the city by the specially appointed professional team.	3 years
3.	The production of the GUP laying down current and future land-use policies in accordance with broader social, economic and political policies.	3 years
4.	Public discussion within residential groups and working organisations; comments forwarded to the Commune Councils.	3 months

5.	The City Executive Council discusses the GUP and considers comments and proposals but forwarded by individual Communes and working organisations of associated labour. When the majority accepts the GUP it becomes the official policy document – otherwise Stages 3,4 and 5 are repealed.	1 month
6.	The revision of the GUP.	5-10 years

Operating the System

After the official acceptance of the GUP, the Implementation Plan (Provedberni Urbanisticki Plan, PUP) is prepared. This consists of long- and medium-term plans and any additional special programmes. Each Commune is responsible for the preparation of three aspects of this document:

(a) a statement of development policy proposals;
(b) urban programmes;
(c) Detailed Urban Plans (DUP), or Master Plans for infrastructure, land-use, etc.

In addition, Commune Councils can:

(a) prepare master plans and projects for minor developments within their jurisdictions (policy section);
(b) issue building permits if existing land-use plans or proposals for future development have already been accepted in principle by the City Executive Council (development control section);
(c) make further proposals for future development. These can be submitted by private, semi-private or public sector bodies, in which case their plans have to be approved by the City Executive Council. These proposals normally consist of information concerning the land-use, density of development, three-dimensional architectural or urban design drawings and the infrastructure plan.

Implementation of particular projects is jointly under the control of the city and the Commune administrative bodies, though smaller projects are solely the responsibility of the individual Communes. It is of vital significance to individual Communes to have their land-use and local master plans approved: otherwise they are not allowed to issue any building permits. The quality of information that each particular Commune produces varies from one area to another, and mainly depends whether they are urban, suburban or rural communities. This information should consist of relevant data on current land-use patterns; statistics related to population projections, transport and infrastructure; and sometimes proposals for future development.

There are shortcomings in this system nearly everywhere, although they manifest themselves in different ways throughout the country. The main problem in almost all communities is a lack of coordination between broader developmental policies and more specific local needs. This is especially evident in the housing sector, where demand continually exceeds supply, regardless of the fact that a large volume of new housing is produced each year.

FORMAL AND INFORMAL SECTORS

Housing is produced in both the formal and the informal sectors. The formal sector comprises housing built by construction cooperatives for both renting (by other cooperatives) or for sale to individuals. The informal sector covers housing built legally by small firms for sale, but consists mainly of illegally built houses.

Formal Sector	Informal Sector	
Cooperatively built flats	Houses built legally by small, private firms.	Houses built illegally by owner-occupiers.
Rented Sold for Owner Occupation Greenfield development, urban renewal		

Suburban Peripheral Growth

Within the formal sector, there are three major types of urban land development:

(a) greenfield sites;
(b) inner-city sites;
(c) urban reconstruction and renewal.

(a) Greenfield Sites
Development of greenfield sites has been the most popular mode of large-scale housing development in post-war Yugoslavia. These developments are mainly associated with the provision of housing by the cooperative sector, which is either rented or purchased for private occupation. In order

to make sure that such large-scale developments could go ahead, city municipalities acquired large sites by compulsory purchase. Owners were compensated financially, on the basis of existing-use value, plus an allowance for disturbance, or with equivalent land elsewhere. Given this basis, the authorities tend to err on the side of generosity, to forestall opposition. Former owners retained a right to use the land up to the time of its development.

In Zagreb, for example, proposals were made in 1953 to develop an area of 2,350 ha south of the River Sava in order to provide accommodation for 250,000 residents. This type of development seemed, at the time, to be the most efficient method of housing provision; it avoided the problems of demolishing urban housing and dealing with complex ownership patterns, and it allowed historic areas to be preserved.

After a number of greenfield sites have been earmarked in individual GUPs, Detailed Housing Programmes are prepared. They are similar in their basic concepts to British 'design guides and briefs', and they generally contain information such as the type, size and location of housing development, density patterns, the relationship between built form and open spaces, the number of houses or flats to be built, type of materials, and so forth. These specifications, however, have to be in accordance with overall housing policies and the requirements laid down in the GUP. Similar policies are also prepared for other types of land use.

Detailed Housing Plans
(DUPs) are a further stage in housing land development. They vary in their concept and presentation, depending upon the type of housing to be provided and the size and location of development. Like urban design briefs, they usually have to be prepared for central and historically sensitive developments, and the development of urban blocks or greenfield sites. They include basic housing layouts for individual sites. They are not required for the development of single plots.

The next state is the preparation of urban design and architectural ideas. The usual process is to invite submissions for a public competition, judged by a jury. Once a scheme has been chosen, tenders are invited from a construction firm. These construction firms are organised on the same basis as other workers' cooperatives (Basic Units of Associated Labour). Large greenfield developments are usually given to several such cooperatives.

When the housing units are completed, they are sold to individual clients or workers' cooperatives, which then let them to their employees. The price is determined by the price of land, construction costs, and local demand and supply conditions. The sale price of a housing unit in large cities is generally about ten times the salary of a full-time worker; prices are lower in small towns and rural areas.

The sale price is also used as a basis for calculating the rent. If a flat is let by a cooperative to one of its employees, the usual practice is that it is 'tied' for ten years, i.e. the employee loses the flat if he changes jobs. But at the end of ten years the occupier is considered to have bought the flat. The private letting of flats by individuals is supposed not to happen.

Tenants pay charges for the cost of maintaining infrastructure and repairing the building. There is also a land tax, based on the area of land divided by the number of dwellings; the rates are higher in central than in peripheral areas. There is also a property tax, but this is at a nominal level.

One of the largest 'greenfield' sites developments is the area of South Zagreb (Novi Zagreb). The original plans were prepared in 1962. The whole area of 2,350 ha, designated for new development, was subdivided into ten separate zones, of which six were designated for residential use. The zones were defined by the layout of the rapid transit system, and each zone was subdivided into residential neighbourhoods, forming a total of 74 residential neighbourhoods. Each neighbourhood was planned to cover an area of 20-30 ha, providing accommodation for about 10,000-12,000 inhabitants. By 1980, all the designated residential zones had been built, and several new ones designated for future urban expansion. Similar 'greenfield' sites were developed in most other large cities.

The greenfield developments have both good and bad points. They have provided large amounts of housing, which is integrated with public transport and social facilities. On the other hand, the layout is often monotonous, and maintenance is often slow and inadequate. Very high blocks have proved unsatisfactory for families with children. The policy now is not to build above five storeys, and to split development and management into smaller units.

(b) Inner City Developments
Most inner city sites have been developed on a piecemeal basis, due to the demolition of older deteriorated buildings. The land of most central areas has been 'nationalised', with the commune as 'ground landlord'; the commune consequently exercises close control over individual developments. Larger sites are developed according to specially prepared design briefs which control the type, size and character of new buildings. When individual sites are developed in peripheral areas, already designated for housing development, building permits are all that is normally required. These sites are usually developed by small scale private builders.

(c) Urban Reconstruction and Renewal

Most Yugoslav cities have only recently undertaken reconstruction and renewal of their deteriorated historic buildings and areas. This has not been due to neglect and ignorance, as almost all larger cities do protect their historic heritage. This is usually done by carefully prepared DUPs (Detailed Urban Plans) which take into consideration both the conservation of older buildings and the gradual renewal of individual vacant sites. The main reason for the lack of adequate renewal is the shortage of finance, especially in situations where buildings are occupied by less wealthy and older residents. Special funds are now being raised (through taxation) in order to save and improve these areas. The cities of Zagreb, Ljubljana, Dubrovnik and Split have been especially successful both in terms of protection and renewal, as they attract additional income from tourism taxes.

Informal Sector Land Development

This type of land development usually takes place in peripheral areas of cities, where there is a lack of master plans to control development, and where there are pressures for urban expansion. The city of Belgrade, which is the national capital as well as the capital of Serbia, can best illustrate this peripheral growth. Recent economic investment in many urban functions (trading, industry, administration, government) has resulted in the expansion of residential areas along major transportation routes. The majority of the city's residents are newcomers from villages and various undeveloped southern and eastern parts of the country. People are attracted to the big city for the fulfilment of their personal aspirations and for an improvement of their living conditions. Unfortunately the city is unable to sustain such a rapid expansion, although a large number of residential estates are constructed each year, through the city's official system of housing provision. Many unhoused families therefore search for alternative housing solutions. A significant proportion of these are self-build houses, built without any building permits. Kalunderica, a settlement of an area of 890 ha, is a typical example of this informal housing provision called divija gradnja (wild building) (Grzan-Butina, 1985).

The settlement pattern of Kalunderica is mainly conditioned by the original village core, and by the size and location of fields and their subsequent subdivision into building plots. Each landowner conducts his own sub-division, and the size of building plots depends upon how much land a farmer wants to sell at any time. It is an area of exclusively private house building, without any officially prepared or approved detailed urban plans. The houses are being built to high construction standards, although without building permits.

The official policy towards 'wild building' has changed over time. The initial method of pulling down illegally built houses did not prove effective, and was soon abandoned. A special document was issued (No. 18/75), which was aimed at preventing further illegal house construction, but with very little effect. In 1981 the settlement had 12,345 residents and, according to unofficial records, about 20,000 in 1985. This type of development is a typical example of rural-to-urban land transformation process, where a traditional village is expanded through informal land subdivision and illegal housing provision. As such, it stands at the margin of rural and urban settlement types.

The land prices in such developments are usually determined by individual landowners and, since the land is classified as agricultural, land taxes are minimal. The construction of individual houses is done by private builders who quite often, in consultation with their clients, follow one of the 'pattern books' which are readily available to anyone in the private housing market. Since these settlement types are not officially designated for housing use, basic infrastructure services and community facilities are lacking.

There are differences of opinion in official circles as to how illegal building should be dealt with. The traditional - and still predominant - view is that it should now be stopped. But the view is increasingly being put forward that it would be better to make provision for more private building, thus legalising it and making it possible to guide development, provide public services and levy taxes. This issue remains to be resolved.

TENURE AND FINANCE

Most cities have a well documented system of land transactions. Every site is entered under a separate number, and plans and details are kept in a document called Katastar. These registers are open to the public for inspection, and must be updated when changes occur. Properties can be purchased either as 'freehold' or 'leasehold', although 'freehold' is more common. All single family, semi-detached and terrace houses are 'freehold'. However, there are situations where the local housing department leases land to potential clients in order to stimulate housing construction in particular parts of cities. These leases are usually for 99 years. Ground rent is paid directly to the ground landlord, i.e. the Commune. The housing built on leasehold land usually consists of blocks of flats let by cooperatives to tenants.

The land which lies within the inner city boundaries is 'nationalised' (or rather, municipalised) and controlled by the Municipal Departments for Urban Development. House owners have a right to use the building but, in the case of any major reconstruction or redevelopment, they are obliged to leave. In return they have a right to an equivalent amount of land at some other location, and financial compensation for lost development opportunities. Municipalities own all land associated with public use such as streets, roads, open spaces, parks, etc.

Type of Finance
There are many ways to finance housing developments. In the cooperative sector, building companies usually invest a part of their resources in the development process; the rest of the money comes from various loans or from the contributions of potential private clients. These clients are either individual families or various workers' cooperatives wishing to purchase houses and flats for their employees. Individual private clients raise their money through personal savings, and mortgages provided by banks. A similar system of finance is used for individual private house development.

There are no national statistics on types of tenure, but it can be said that owner-occupation ranges from 30 to 60 per cent in different cities. In agriculture, owner-occupation is fairly universal. Rents are subsidised in various ways, and depend on household income and the cost of construction. They are also influenced by the local demand and supply systems, and vary according to city and location.

CONCLUSION AND PROSPECT

The current land development system in Yugoslavia is deeply rooted in the present administrative and political structure of the country. Municipal town planning bodies are largely responsible for the promotion and control of land development, through a well-established hierarchy of local and regional plans, conservation, urban renewal and inner city redevelopment are carefully coordinated through Detailed Urban Plans. Most large-scale housing developments have taken place on 'greenfield' sites. Although a large number of housing units are constructed every year, it does not meet the ever-increasing demand. Many self-build housing schemes, associated with the urbanisation of peripheral areas, have been established as an alternative of the official system. As a consequence, there is now a wide range of housing types available. Finance is provided by the cooperative sector and by individual investments supported by bank loans. It is likely that this dual system of land development will continue for some time, although more emphasis should be placed on urban renewal and improvement of older buildings in order to preserve a valuable historical heritage.

REFERENCES

Auty, P. (1965) 'Yugoslavia', Walker and Company, New York

Grzan-Butina, G. (1982) 'The Contemporary Town Planning System in Yugoslavia', unpublished paper, Oxford Polytechnic

Grzan-Butina, G. (1983) 'The Pattern of Modern Yugoslav Cities: A Comparative Study of Physical Form and Spatial Structure in Zagreb and Ljubljana', PhD thesis, Oxford Polytechnic

Grzan-Butina, G. (1985) 'Kaluderica: Observations on a Wild Settlement', unpublished Field Study Report, Oxford Polytechnic

Peric, L.J. (1985) 'Development of Kaluderica', unpublished Diploma Study, University of Belgrade

Chapter Seven

GREAT BRITAIN

Graham Hallett and Richard Williams

INTRODUCTION

The most striking characteristic of British land policy since
1945 has been its violent swings. Three controversial tax-
ation/land buying schemes have been introduced by Labour
Governments, and repealed in their entirety by Conservative
Governments. Since 1979, an equally controversial 'radical
Conservative' programme has been progressively implemented.
During this post-War game of 'political football', the possibilities of developing forms of taxation and state intervention
designed to correct and supplement the land market have
largely been sacrificed. In other countries there have been
changes of course, but nothing comparable to the series of
'U-turns' which the British ship of state has executed. The
bulk of this chapter is therefore devoted to examining the
extent to which land policy has been subject to political
vicissitudes, and the implications for other sectors of policy -
housing, land-use planning and economic development - with
which land policy is intimately connected. After some background comments on 'the British constitution' and the historical background, we shall summarise post-War land policy
developments, before and after the watershed of 1979, and
then examine current issues concerning land prices, town
planning, land acquisition, Urban Development Corporations
and 'rates' (property tax).

In the chequered history of British land policy, a domi-
nant role has been played by 'instant legislation'. Policies in
this field (and others) lack a basis of political consensus, and
are not regulated by a legislative 'flywheel'. Legislation can
be changed virtually overnight if there is a change in the
political complexion of the central Government. This situation
reflects both political polarisation and the peculiarities of the
British constitution. There is no written constitution; no
legislature independent of the executive; no federal system;
no proportional representation. The House of Commons today
functions more as an electoral college for the election of the

114

presidential Prime Minister than as a legislature. The House of Lords is a more competent legislative body, but has no power; successive Governments have retained its outdated hereditary membership in order to justify the virtual absence of a second chamber. If a Party has a majority in the House of Commons (even if, like all the Thatcher Governments, it has a minority of votes cast) it feels itself to have a 'mandate' to introduce any legislation, however controversial, and it can have an Act passed within a few months. No other country in our study has such an absence of 'checks and balances' on the power of the central Government.

Political polarisation extends to local as well as national government. According to a formerly eminent town planner;

'In the 1960s and earlier, planning was dominated by paid officials, and politics at the local level was mainly concerned with adjusting the steering rather than changing the direction. Now we have a much more politicised system and elected members rather than officers very often seem to make the running on planning matters. The continuity of thinking between officers and local politicians (irrespective of party) that resulted in the rebuilding of Coventry or Plymouth, or later in Newcastle upon Tyne, no longer exists to any significant extent' (Burns, 1983).

But there are other weaknesses in local government, which were apparent before 1979 (Eversley, 1974; Newton, 1981; Loughlin, 1985a). One was that local authorities controlled only one, notoriously inelastic, source of revenue - the property tax, or 'rates'. Local authorities therefore became heavily dependent on central government grants, with strings attached. Since 1979, however, the process has been accelerated by a Government explicitly hostile to local autonomy. Local government has steadily been emasculated, and the Metropolitan Counties responsible for Greater London and the other large conurbations have been abolished. London - alone among 'primate cities' in developed countries (Norton, 1983) - now has no elected governing body, and the world famous County Hall is up for sale.

THE HISTORICAL BACKGROUND

Much of the post-1945 legislation could be said to be fighting the battles of the 19th century. At that time, Britain had - by West European standards - an unusually concentrated system of landownership. Most farmland was in large 'estates', whose aristocratic owners let land to tenant farmers, who in turn employed over a million labourers. In some cities, there were also large leasehold housing estates. Leases were usually of 99 years. At the end of the lease, the land and house reverted to the ground landlord, without any compensation for the house (which seemed inequitable to everyone except ground landlords and land lawyers). The leasehold estates embraced the London 'Belgravia' estate of the Duke of Westminster which was an affluent, well managed, area; the working-class housing in South Wales, where the ground landlords were mineowners who performed no management functions at all; and the Bournville Estate in Birmingham, which pioneered the 'garden village'. The leasehold system was thus not all bad, but 'The Land Question' was given a particular bitterness by the existence, in both town and country, of large private landholdings owned by a privileged class (Douglas, 1976).

In addition to the distinction between leasehold and freehold tenure, there was the distinction between owner-occupied housing and tenanted housing. The holder of a long lease, or a freeholder, could either occupy the dwelling himself or let it to a tenant on a weekly basis. The leasehold system was confined to certain districts, most housing being freehold. On the other hand, around 90 per cent of housing was tenanted. In this respect, Britain was not very different from continental Europe. It was, however, different in that - except in parts of London and Glasgow - most housing consisted of terraced houses rather than blocks of flats.

The two World Wars, and the social and fiscal changes they brought, led to the decline of both the rural and the urban estates, and the private tenancy system. There has been a swing to owner-occupancy, which now accounts for over 50 per cent of agricultural holdings and 62 per cent of dwellings. The old urban leasehold system was brought to an end by the Leasehold Reform Act, 1967, which extended the life of existing leases, and gave leaseholders a right to buy the freehold. This change was naturally very popular with leaseholders, if not with ground landlords. Some land economists questioned the wisdom of fragmenting the estates, rather than transferring freehold ownership to local authorities or other public bodies, so as to obtain some public benefit from rises in land values and make possible legal control (beyond that available under town planning law) over maintenance standards and modifications to houses; such views, however, had no influence at all.

Before 1914, most local authorities intervened relatively little in the land market. As a Liberal Party report in 1925 concluded;

'... it is abundantly evident how far this country lags behind many of its Continental neighbours in regard to public acquisition and control of land ... The Local Authorities of Norway or Holland or Germany have a freedom undreamt of by us in regard to this question' (Liberal Party, 1925).

The report recommended improving, rather than abolishing, the leasehold system. It further recommended site value taxation; a betterment levy where land values had been increased by specific public improvements in the vicinity; and the expansion of municipal landownership 'not by sudden and indiscriminate purchase, but by a judicious policy of steady acquisition in advance of actual development'. It differed from the Labour Party (which was committed to nationalising all land) in supporting, alongside municipal intervention, 'the free operation of private enterprise' and 'private and individual tenure of land'. When, in the 1930s, the Labour Party replaced the Liberal Party as the Opposition to the Conservative Party, this type of 'mixed economy' land policy ceased to be represented by either of the two main parties.

It should not be thought, however, that all local authorities adopted a laisser faire attitude before 1939. Indeed, in some respects they used land acquisition - notably for the preservation of open space - more effectively than they have done since. In the inter-War period, some authorities took advantage of low land prices (and the financial difficulties of the landed gentry) to buy 'country houses' with parks, or stretches of woodland, on the urban fringe. Nearly all the most attractive park areas in and around British cities were acquired in the supposedly 'dark ages' before 1947.

Inter-War Housing

The inter-War period saw a vast expansion of unplanned suburban housing for sale (often 'ribbon development') and of 'council housing'. Government subsidies were provided so that houses could be built for letting at low rents to people on low incomes. These subsidies were provided solely to local authorities, who were given powers to buy land compulsorily, organise the building of dwellings, and let and manage the houses themselves. In this way, local authorities eventually came to own and administer one third of the housing stock, usually in the form of large 'council estates', which were geographically and architecturally distinct from other areas.

Not until 1967 were subsidies provided for housing associations, and the housing association (or 'non-profit' or 'voluntary') movement has not yet made up for the head-start enjoyed by the councils. The subsidy system introduced in 1918 - combined with the undermining of private tenancy by rent control from 1939 to the present day, and the tax concessions for owner-occupancy since the 1960s - put Britain on a path which led to the polarisation of housing between 'council estates' and owner-occupied areas, with a very small 'housing association' sector, and a declining private rented sector, operating largely outside the law (Edinburgh, 1985).

The Taxation of Owner-occupation

Until the early 1960s, Britain had a system for taxing owner-occupied housing which might have been devised by a philosopher-king. Tax relief on (all) interest payments was balanced by a 'Schedule A' tax on imputed rent, based on rateable value. The tax system was thus neutral as between tenancy and owner-occupation. Maintenance expenditure could, moreover, be set off against the tax, thus providing an incentive for owner-occupiers to maintain their homes. As the number of owner-occupiers rose, however, the abolition of the Schedule A tax was seen as a vote-catcher, and it was abolished. Tax relief for interest payments in general was also abolished, but an exception was made in the case of mortgage loans on owner-occupied dwellings (up to a maximum loan of, currently, £30,000; £60,000 for an unmarried couple!). The system has many anomolies, one of which is that much of the tax relief on 'mortgages' finds it way into non-housing expenditure. Some critics seem to regard owner-occupancy as the source of all evil (Ball, 1985). Even those who do not take this view can have reservations about a system which gives the greatest assistance to those with the most expensive houses and the highest incomes.

THE POST-WAR PERIOD

The developments since 1945 can be examined under the headings of;

1. revolutions and counter-revolutions concerning the taxation of development value,
2. town planning and urban renewal before 1979,
3. the 'Thatcher experiment'.

THE TAXATION OF DEVELOPMENT VALUE

This is a strange tale, but an instructive one. If our necessarily brief summary taxes the reader's credulity, we would refer him or her to fuller accounts (Prest, 1981; Cullingworth, 1981; Hallett, 1977).

The 1947 Town and Country Planning Act introduced a new system of town planning. All 'development' requires 'planning permission' from the local authority. 'Development' is very broadly defined to include not only building, but also 'material change in the use of buildings or other land'. If consent is refused, the applicant can appeal to the national

118

Minister (the Secretary of State for the Environment). In most cases, of course, officials make the decisions, on general principles laid down by the Secretary of State. When planning permission is refused, there is no compensation. This system has remained. The Act's financial provisions, however, were less durable.

All planning permissions (after 1947) were accompanied by a 'development charge'. This was a tax equal to the assessed 'development value', i.e. the gain from development. This tax, and subsequent ones, were sometimes referred to as 'betterment levies' but this was misleading. The taxes were on development, and on all development. They were thus different from the charges which had been levied, in the 19th century and earlier, on local landowners who benefited from public works such as flood protection (Hallett, 1977, p. 124).

The 'development charge' (if correctly calculated) removed all profit from the land aspect of development. For example, someone selling off part of a large garden for housebuilding would receive no compensation for the loss of amenity, and would be disinclined to do so. The system actually meant that it was impossible to rely on the market mechanism, which is based on the pursuit of profit, for the supply of building land. The corollary to the development charge would have been a system of compulsory purchase of all land which was to be developed, but such a system was not introduced.

During the Labour Governments (1945-51), however, the development charge had little effect, since little private housebuilding was permitted. Most housing was council housing, on land acquired by compulsory purchase. But as private housebuilding began to revive in the 1950s, the deterrent effects of a 100 per cent tax on development value began to be clear. One of the problems (which recurred with the subsequent taxes) was the difficulty of assessing the charge when land was not actually sold. For example, if a developer built an office on land which he owned, and retained ownership, he would incur a development charge based on the capitalised value of future profits. There was a large margin of error, and tax had to be paid on income which had not yet been received.

The development charge was abolished by the incoming Conservative Government in 1951. As there was at the time no capital gains tax or wealth tax, and no provision for charging income tax on land sales by individuals, this meant that gains from land sales or development were free of tax, a situation widely considered inequitable. When the Labour Party returned to power in 1964, it modified the earlier plan. It proposed setting up a Land Commission with wide powers to acquire land compulsorily (DOE, 1965). This idea found support in town planning circles on the grounds that the purely 'negative' planning exercised through development

119

control, needed to be complemented by 'positive planning' involving land acquisition. (Which was neither a novel nor a particularly socialistic concept, since it had been introduced by Konrad Adenauer in Cologne forty-five years previously; see the German chapter). The Commission would also administer a 'betterment levy', which was similar to the 'development charge', but differed in two respects. Firstly, it would take only (initially) 40 per cent of development value. Secondly, it would be levied not only at the time of development, but on various other occasions, such as a sale, or the granting of a lease.

The White Paper was hailed by one 'centrist' economist as, 'a scheme whose main outlines can hardly be rejected by anyone who wants to try to combine economic efficiency with social equity' (quoted in Hallett, 1977, p. 129). Indeed, the principle of a 40 per cent tax on substantial development value, and a land-buying agency, could well have been the basis for a workable scheme. In the event, the 1965 Land Commission Act can lay claim to be the most complicated and obscure on the British statute book. The levy became an administrative nightmare, because of its Byzantine complexity, and the way it was applied to the smallest or most notional of gains. The Land Commission also faced political problems. The Chairman of the Commission (taking a refreshingly different line from the politicians who set it up) saw it as,

> 'a body which will introduce into the whole planning machine what you might call a sense of commercial responsibility ... If you get planners working entirely on their own, without continually being reminded that they are using national resources, they sometimes go absolutely haywire' (Economist, 1 April 1967).

The Commission was, not surprisingly, viewed with suspicion by local planners, and incurred odium among the public because it administered a levy which sometimes caused hardship. In 1969, the Government realised that the levy was spreading its net too widely, and exempted some small transactions. Further steps in this direction could probably have made the levy acceptable. In the following year, however, the Conservatives returned to power, and abolished both the levy and the Commission.

The period 1972-73 saw an unprecedented property boom (Figure 7.1). Prices of land, houses and offices soared and there was a public outcry against 'speculators'. It is clear enough in retrospect that the boom was unsustainable, and that it was caused by a deregulation of the banking system - which allowed the mushroom growth of some dubious (and short-lived) banks - and by one of the periodic bouts of hysteria to which markets are prone (Mayes, 1979). The boom

Fig. 7.1:   Average housing land prices for England and Wales
1970-85, in relation to (a) building prices,
(b) average income.

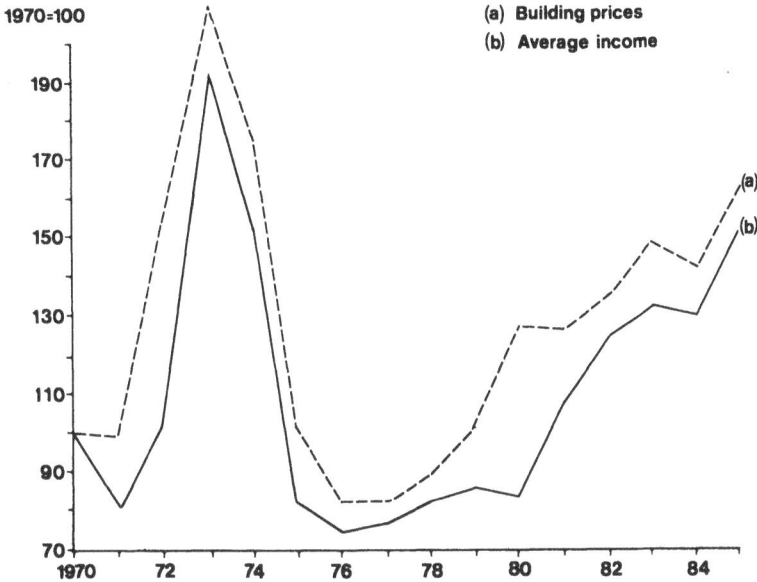

was quickly followed by a slump, but not before it had a marked effect on policy.

In Opposition, the Labour Party moved sharply to the Left, and adopted a new policy of, in effect, land national-isation (DOE, 1974). In 1975, the new Labour Government passed the Community Land Act. The idea was that (after a transitional period) all land required for development or redevelopment would be compulsorily acquired by local authorities, at existing use value. Developers would have to turn to local authorities for land, which would be disposed of at current use value. During the transitional period, there would be a Development Land Tax (similar to the 'betterment levy') at a rate of 80 per cent.

The problems to which this legislation would give rise were predicted in a report by the Royal Institution of Chartered Surveyors (RICS, 1974). Of the 'ultimate situ-ation', it stated;

'This concept pre-supposes that control and development of land can be achieved only through land being in public ownership. In our view there are four prerequisites for successful development: the will to develop the professional and technical expertise, the financial resources and the availability of suitable land. The mere ownership of land cannot by itself cause development to take place' (RICS, 1974).

The first sentence correctly interpreted the Government's philosophy, and the second and the third sentences find substantial support in the history of a great deal of publicly-owned land. On the other hand, there was a case for arguing that some new mechanism for the public acquisition of land in advance of development - alongside the 'strategic weapon' of compulsory purchase - was necessary. A subsequent report concluded that the Act was not fulfilling its stated objectives, and was even less likely to do so in its ultimate form (RICS, 1978). The report examined two possibilities:

(a) to repeal the Act and encourage 'positive' planning by new legislation giving local authorities greater compulsory purchase powers and the right to impose infrastructure charges,
(b) to amend the Act, with the same aim.

Some 'liberals' hoped that the new Conservative Government of 1979 would indeed amend rather than abolish the Act, and end the disruptive cycle of legislation and repeal. Such an approach, however, was alien to the re-organised 'Thatcherite' Conservative Party. The new Government immediately repealed the Act, as it affected local authority buying powers. The Development Land Tax - with which the commercial developers had learned to live happily enough, but which could, in any case, have been improved in detail (Prest, 1981) - was repealed later.

Thus nothing remains today of an enormous expenditure of time and money - with one exception. In framing the Community Land Act, the Labour Government decided that, in Wales, the functions entrusted elsewhere to local authorities should be undertaken by a nominated regional body, the Land Authority for Wales. (According to some experts, the reason was the nepotism of Welsh local authorities.) The Authority just managed to survive the repeal of the Act, and today provides a useful, uncontroversial land acquisition service for local authorities in Wales.

At the present time, the saga we have outlined above tends to be either dismissed as ancient history or judged in (differing) black-and-white terms. Our assessment is that more will eventually be heard of the issues involved, but that, although the schemes should have been reformed rather than repealed, they were flawed by a confusion of objectives. If they had had the feasible objective of setting up land-buying agencies linked to local authorities, and imposing taxation of, say, 30-40 per cent on development gains, they could have been workable and beneficial. But they were based on inadequately thought-out doctrines of 'taking (all) the profit out of development'. The 1965 scheme was, in principle (as distinct from detail), the most workable, but it suffered from not being linked to local authorities. In all three cases, there was a failure of the legislative system. To get a draft law right would have needed several years of discussion, informed comment and amendment; instead, the Labour laws were drafted virtually without consultation, and passed in months rather than years: so were the Conservative repealing laws.

Planning Gain

The whole emphasis of the Labour Governments was on preventing landowners benefiting from 'unearned increment' by means of national taxes; the alternative approach of financing the cost of public infrastructure through levies on developers - and hence ultimately landowners - was neglected. Thus no system of 'infrastructure charges' was initiated. The cost of infrastructure had to be borne by the public utilities or the local authority. More recently, a loosely defined procedure known as 'planning gain' has grown up. Under the 1971 Town and Country Planning Act (Section 52) a local authority may make an agreement with a developer requiring him to undertake certain works. These agreements may thus make the granting of planning permission dependent on bearing the cost of providing sewers, or of building a community hall, or even of making a cash contribution to some worthy local cause.

The term 'planning gain' thus covers a multitude of possible sins. It can be argued that the procedure is a legitimate way of extracting 'unearned increment' - especially in the absence a formal alternative (Keogh, 1985). The courts, however, have taken a very narrow and individualistic approach to the question, which has been criticised as inappropriate to large-scale urban development (Loughlin, 1985b). A report by an advisory group came down against a general adoption of 'planning gain' but considered it legitimate for a planning authority to require the developer to undertake essential infrastructure, or to impose conditions when certain aspects of the scheme were objectionable (DOE, 1981a). The Committee's cautious approach was probably influenced by the fear that a 'free for all' would leave developers open to the most rapacious demands. There is still no clear-cut law on the matter.

GREAT BRITAIN

TOWN PLANNING AND URBAN RENEWAL BEFORE 1979

New Towns

During all the political battles on the taxation of development value, a lot of housing was built, under the planning framework of the 1947 Town and Country Planning Act and the New Towns Act, 1946. The New Towns attracted particular interest in other countries. The idea was taken over by the State from the private 'garden city' associations which originally built Letchworth and Welwyn, as a result of the ideas of Ebenezer Howard (Howard, 1898, 1945). The aim was to build a number of self-contained towns in open country. It was hoped that in this way the excessive congestion of steadily-growing large cities - especially London - would be avoided.

A New Town Development Corporation was appointed for each planned town, with special statutory powers, including the power to purchase land compulsorily, at existing use value. An appropriate area of land was then designated and acquired compulsorily, most of it at agricultural value. The Development Corporation organised the construction of the town, disposing of commercial and industrial land on long leases and building housing for rent. In the early days, houses were allocated only to persons who obtained work in the town. The earlier New Towns were of 50,000-100,000 inhabitants, and monocentric. Different types were subsequently developed, such as the 'growth pole' (Washington), the 'expanded city' (Peterborough) or the 'new city' (Central Lancashire, or Milton Keynes) which absorbed existing small towns.

The New Towns were at first hailed as the New Jerusalem, but there then followed a reaction, in which their dullness ('New Town blues') and often rather shoddy architecture was emphasised (Evans, 1972; Aldridge, 1979). It now seems possible to make a more balanced assessment (Cullingworth, 1985). The earlier New Towns suffered from uniformity and drabness, but they are improving with age. Milton Keynes, the last New Town, began well, and has many virtues, including a million newly-planted trees. However, as housing subsidies were cut, and the Corporation's power restricted, standardised commercial development of owner-occupied housing became predominant. Milton Keynes has also, like most British cities, suffered from 'de-industrialisation'. But even though they may not have lived up to excessive initial expectations, and may sometimes have been 'over-planned', the New Towns were impressive achievements. They showed that, with the help of appropriate land policy, it was possible to provide housing and employment in a community of manageable size, and avoid the disadvantages of peripheral growth around the large metropolises.

The New Town Development Corporations are now being wound up, and their assets being sold off; the planning ideals represented by the New Towns do not at present enjoy favour in the corridors of power. However, one aspect of the constitution of the New Town Development Corporations - that they were nominated, and displaced local authorities - has been revived in the Urban Development Corporations, set up to tackle inner-city regeneration.

Urban Renewal

In the 1950s and 1960s, Britain not only built New Towns, but engaged in 'slum clearance' on an unparalleled scale. Huge areas of 19th century housing were bought up by local councils and bulldozed. Redevelopment took rather longer, and some cleared sites remain derelict to this day. Moreover, the redevelopment which did take place often proved unsatisfactory. It consisted solely of council housing, and the lay-outs and construction methods used (encouraged and sub-sidised by the central Government) were frequently defective. There were special subsidies for high buildings, and the 'deck access', walkways, and other fashions of the time provided an ideal environment for vandals and thieves (Coleman, 1985). Moreover, 'system built' housing experienced problems of insulation and condensation, and soon began to display severe structural faults; the (partial) collapse of Ronan Point in 1967 symbolised the beginning of the end of an era. (Only recently, however, was it demolished, with explosives.) The cost of repairs and replacement for this type of housing is currently put at £30bn, but local authorities have been starved of money, and even prevented by central Government from spending the receipts from selling council houses.

The defects of the new housing were due to the naivety (not to use a harsher word) of the architects and planners, but this naivety was allowed full rein by the organisational system. A single political body planned large areas of housing, built it, selected the tenants, let the dwellings to them, and managed the dwellings. This system combined the dangers of 'all eggs in one basket' with those of excessively large organisations and politicised decision-making.

Compensation for Compulsory Purchase

Another factor in the doubtful success of 'urban renewal' was the compensation code for housing acquired compulsorily (Hallett, 1979). Various Acts of Parliament provide the basis for public bodies - both local authorities (subject to approval from the central Government) and nationalised industries, to acquire property compulsorily, on payment of compensation. The general rules for compensation were laid down in the

Land Compensation Act, 1919, and have survived the post-1945 legislative see-saw. The principle is that compensation is at 'market value', subject to two qualifications. Firstly, 'market value' means the price that a willing seller might reasonably be expected to realise on the open market, ignoring any unusually high price which one buyer might offer because of special circumstances. Secondly, any increase in value arising from the acquiring authority's own development is ignored. The basis for compensation is thus a circumscribed 'market value', with limited provision for recouping 'betterment'. There was an exception, however, in the case of housing declared to be 'unfit'. Such housing could be acquired without compensation; the only payment was for the value of the site. This expropriation without compensation might have been reasonable if 'unfit' had been used in its original sense of a danger to health. But, in the 1950s and 1960s, new standards based on current building practice were adopted; as a result some old houses, which were quite habitable, were compulsorily acquired for less than a tenth of their market value. (This ruling is still applied to tenanted housing, although it has been relaxed in the case of owner-occupied housing.) A compensation code which restricted the definition of 'unfitness' to its original meaning, and therefore limited expropriation without compensation to hovels, would probably have encouraged a more humane and economic type of urban renewal, with more limited demolition and more encouragement for the maintenance of older housing.

The Swing to Rehabilitation
In the 1970s there was a swing to rehabilitation and, in new construction, traditional housing styles and conventional methods of construction. In both respects, the pendulum may have swung too far. Moreover, the rehabilitation of older areas has often been of a 'pepperpot' (i.e. scattered) kind, with serious dereliction in between. 'Pepperpotting' has been encouraged by the payment of rehabilitation grants to individual owners, as compared to 'area' improvements. There have been area schemes (Housing Action Areas) but they have rarely been large enough, or comprehensive enough, to make a major impact on the problem.

The early urban renewal plans were purely architectural in their approach; only in the 1970s were social issues investigated. Studies of 'inner areas' in London (Lambeth), Liverpool (Toxteth) and Birmingham (Smallheath) were sponsored by the Department of the Environment (DOE, 1977b). They made it clear that British cities were experiencing what had previously happened in American 'central cities'; the concentration of the poor, the unskilled and immigrants; the decay of old housing; the decline of business activity; the problems of crime and poor schools. In addition, there was the suffering caused by 'the municipal bulldozer', combined with neglect by the city authorities in other respects. The reports recommended measures to increase employment and improve education and training in the inner areas, alleviate poverty more effectively, decentralise the administration of council housing and lay more emphasis on housing associations. The Labour Government responded with a White Paper (DOE, 1977a), the main proposals of which were to ease planning controls on small industrial premises and to set up 'partnerships' with local authorities in the main conurbations, whereby extra funds would be channelled to 'designated areas' under the 'Urban Programme'.

'Partnership'
A different type of 'partnership' involved the use of agencies, such as the Development Agencies which were set up to create employment in Scotland and Wales. In 1976 a programme was initiated in the East End of Glasgow to renew what was described at the time as the most deprived city district in Europe. This took the form of a partnership between the Government, the Scottish Development Agency (in the coordinating role), the Scottish Special Housing Association, Strathclyde Regional Council, Glasgow City Council, and some specialised agencies. The partnership worked quite well, and (with the aid of considerable funds) has certainly effected a transformation of the housing in the area (Wannop, 1985).

However, even these limited interventions by central Government reflected growing doubts as to whether local government could deal adequately with the problem of 'inner areas'. There were certainly weaknesses in local government, which were increased by its 'reform' in 1972. Up till the 1970s, local government had broadly maintained the structure which had been set up in the mid-19th century. The reorganisation suffered from the familiar 'instant legislation', and from the fact that it was confined to the size of authorities, ignoring the far more pressing problem of finance. The incoming Conservative Party in 1970 - abandoning a plan which had been prepared by its predecessor - adopted a two-tier system of 'districts' (which can be large cities) and 'counties' (in Scotland, 'regions') which soon proved to be a mistake. The change itself involved enormous dislocation, and the division of responsibilities between the two tiers was obscure and unsatisfactory. No change is likely, however, because no one wishes to go through another reorganisation.

Local Authority Land Policy
We have indicated the large swings in national policy under which local authorities had to work between the 1940s and the

127

1970s. The way in which local authorities operated 'land policy' (in the sense of purchases and sales) over this period is not well documented. One of the few available case-studies (of Oxford and Sheffield) is resolutely Marxist in tone, but contains useful information (Montgomery, 1987).

It first needs to be explained that local authorities can buy and sell land like a private company - if they have the money - or they can use compulsory purchase. The latter, however, is legally circumscribed, and requires the consent of the central Government. Land can (or could) be compulsorily purchased for the provision of council housing. Compulsory powers can also be used in conjunction with private development, e.g. a council may buy up land for a town centre development, to be carried out by a private developer. When the development is completed, the buildings can be sold, on either a freehold or a leasehold basis. The choice will depend on both the demand by potential purchasers, and on central Government policy. Under the 1975 Community Land Act, local authorities were encouraged to dispose of land on a leasehold basis; this is not favoured by the present government.

Montgomery's study of Oxford and Sheffield shows that both cities acquired considerable quantities of land in the period 1945-55. This acquisition was assisted by the ability to acquire property at existing use value, and 'unfit' housing at site value. Montgomery criticises the influence of commercial interests in the redevelopment of a 'slum' area in Oxford, and in the acquisition of land for the steel industry in Sheffield; with the decline of the Sheffield steel industry, the land remained unused for years. But pressures from industry are not unknown in countries where all land is state-owned, and it seems hard to blame Sheffield Council for not foreseeing the near-demise of its principal industry, and acting on that assumption. One interesting point brought out by Montgomery is that the Community Land Act, which was supposed to expand the local authority's role in the land market, was so badly drafted that it actually inhibited it.

In the 1970s, the role of local authorities (and New Town Corporations) in the land market began to decline, and this decline has accelerated in the 1980s, as a result of financial pressure from the central Government. Local authorities still collaborate with private developers in the land acquisition process, but virtually no local authorities would now be in a financial position to undertake 'land banking', or be allowed to do so. The private sector has to some extent expanded to fill the gap. Housing development (for sale) has become concentrated in the hands of about a dozen large private housebuilding firms, who negotiate with private landowners - both on greenfield sites and increasingly within built-up areas - for land acquisition. In a similar way, forms have grown up which specialise in the construction of out-of-town shopping centres and other commercial developments, handling the entire process from the purchase of land to the disposal of the buildings. Except for the expanding activities of the Urban Development Corporations, the system of land acquisition has thus become increasingly 'American'.

THE THATCHER EXPERIMENT

The 1970s saw the onset of profound changes in the structure of the economy, which were intensified in the 1980s. These changes have altered the nature of land-use planning.

'(Before 1969) the planning system was concerned with ridding the country of the worst legacies of the industrial revolution ... Today the Victorian industries that were the life-blood of the inner areas of our cities have disappeared ... Manufacturing, which is in any event declining in terms of numbers employed, now wants to be sited in the suburbs and increasingly in small towns and rural areas; the information and high technology industries needing office or mixed office and manufacturing space have expanded enormously; the communication systems have been modernised and expanded and now give enormous flexibility to those contemplating development. And hanging over us all today is the fact that we have changed from a growth economy to a stagnant economy and no-one sees how we are going to get out of that situation' (Burns, 1983).

To the underlying structural changes was added the 'Thatcher experiment'. Mrs Thatcher explicitly rejected what remained of the post-War 'consensus'; her aim was to destroy the 'socialism' which had developed over the previous century and establish (or re-establish) an 'enterprise economy' based on private ownership and the market (Riddell, 1985). In the field of land policy, the main changes have been the repeal of the Community Land Act, the weakening of local development control, the selling-off of council housing (through giving tenants a 'right to buy' at discounts of up to 70 per cent on market value, and giving local authorities the right to sell their estates to private developers), the creation of Urban Development Corporations and Enterprise Zones, and the abolition of the 'rates' (property tax).

Privatisation and deregulation, however, have been combined with political and administrative centralisation (Burgess and Travers, 1980). Legislation has been passed enabling the central Government to decide how much each local authority may spend, and how much it may raise from the rates. In housing and land policy, moreover, (as in education, policing and other matters) local authorities are increasingly being by-passed.

There are virtually no official documents explaining the rationale of this centralisation policy, and one has to rely on asides by Ministers, or press comments. Ministers have denounced the 'prodigal spending' of Labour-controlled local authorities, and argued that a Government has no control over the economy unless it can determine the spending of individual local authorities – a view which would be found surprising in the other countries studied, and is challenged by academic specialists (Travers, 1987; Jones and Stewart, 1985). A more general criticism is that Labour local authorities are 'loony', 'totalitarian', and that 'they stink', to quote Her Majesty's Ministers. Another recurrent argument is that Labour authorities have driven away business through the high level of business rates (the annual tax on business premises); the only available study of the subject concludes, however, that rates are not a major factor in the location of businesses (Crawford et al., 1985). A further argument concerns the alleged inefficiency of democratic local government; the 'radical, free market' Economist contains articles with headlines like 'Centralisation good, decentralisation bad', and 'Cut the democracy, cut the delay' (11 April 1987).

Unemployment and Poverty

The background to all this is an exceptionally high level of unemployment. The period 1979-82 saw a loss of employment which is unprecedented in British 20th century history, and unparalleled among major industrial economies. A sharp regional division has also developed, with the highest regional unemployment rates in the North and West of the UK. Moreover, inequalities of income – which had fallen over the post-War period – have increased sharply. The real income of the top fifth of households rose by 22 per cent between 1979 and 1985, while that of the bottom decile fell by 9.7 per cent (CPAG, 1987).

One consequence of a more unequal distribution of income is that, while there are large numbers of buyers for homes at six-figure prices, many people have difficulties in paying for housing – or even finding rented housing; the private rented sector has continued to decline, and a million council homes have been sold to tenants. Council housing is becoming a 'last resort' for the poorest section of the community. The numbers of persons accepted as 'homeless', the numbers of families in 'bed and breakfast' accommodation, the numbers on council waiting lists, and the number of mortgage repossessions have all increased several-fold. It should be added that these are minority problems; most families are not poor, not unemployed, and not badly housed. Indeed, one survey found that in 1986, only 5 per cent of adults were 'dissatisfied' with their housing (Boleat, 1986). But the 'housing problem' in developed countries today is a minority problem, and in Britain the size of the minority (on many definitions), and the intensity of its problems, has increased in the 1980s.

Infrastructural Decay

To the growing 'housing problem' has been added a deterioration in the nation's physical base. The cut-backs in public expenditure (which began under the Labour Government in 1976, but accelerated under the Conservative Government) have been concentrated on housing and 'infrastructure' - roads, sewers, bridges etc. The results are increasingly apparent; one study concludes,

> 'The infrastructure, ill cared for and vulnerable in 1980, is now (1985) sufficiently unkempt and at risk to give rise to public alarm' (Manser, 1985).

The Inner Areas

National trends have been magnified in the 'inner areas' and the large council estates, where unemployment and poverty are concentrated; unemployment rates are often over 50 per cent (Economist, 1982). These districts erupted into violence in the 1980s. There were serious riots in London, Birmingham and Liverpool in 1981, and again in 1985 and 1986. After the 1981 riots, the Government continued - and indeed increased - the Urban Programme of assistance to 'big city' authorities, but it was also determined to curb what it saw as the excessive spending by the same authorities. It opposed, for example, the large council-house building programme of Liverpool City Council. The Government thus gave with one hand and took away with the other, so that on balance less finance was directed to the cities with the gravest problems.

URBAN DEVELOPMENT CORPORATIONS

When the 'inner areas' once again erupted into violence in 1985, Government Ministers at first tended to treat the riots as, 'Not a cry for help, but a cry for loot' in the words of the current Secretary of State for the Environment. He went on to argue to that the problem could not be solved 'by throwing money at it', and was indeed largely the result of past expenditure by local authorities. On reflection, however, the Government decided that it would throw a certain amount of money at the problem - but not through the detested Labour-controlled local authorities.

The thinking behind what has become the cornerstone of Government policy towards the inner areas is reflected in a passage from the Sunday Times in 1981, quoted approvingly by a stern 'free enterprise' critic of local planning (Moor, 1985).

'Some people in Westminster and Whitehall (i.e. central government and the civil service), are now beginning to doubt whether any funds can be properly directed under Britain's current local government structure. To try to circumvent this kind of obstacle, one radical idea being discussed is that new types of city development corporations, funded by central government, banks and insurance companies should be imposed on local councils. The task of these corporations would be to haul the inner areas up to national average standards of employment, housing, and so on ... The difficulties of "taking into care" for perhaps 10-15 years are obvious, but some experts now believe that only such drastic action can work'.

'Taking into care' became official Government policy in 1985, when it was announced that five Urban Development Corporations (Trafford, Tyne & Wear, Teesside, Black Country, Cardiff Bay) were to be set up, to undertake development in specified urban areas, by-passing local authorities altogether - and massively financed from the central Government.

The forerunners of the new UDCs were the London Docklands Development Corporation (LDDC) and the Merseyside Development Corporation, set up in 1981. Until the 1950s, the central London docks constituted the largest British port. But as trade moved down-river, the docks gradually closed, leaving a derelict area of some eight square miles of land and water. At first, redevelopment was frustrated by an unwillingness to recognise that the docks were in irreversible decline; by shortages of cash and poor communications; and by disputes between local authorities. In 1981, the then Secretary of State (Mr Heseltine) handed over the whole area to a special nominated body, the LDDC, which - in marked contrast to the local authorities - was amply supplied with government grants and loans. Within the area, the Corporation took over the planning functions of a local authority, and became responsible for land disposals. It has sold considerable areas of land to housebuilding firms, and has attracted a number of businesses, especially printing works. A huge office complex, Canary Wharf, is planned for the 'Isle of Dogs' Enterprise Zone, within the area. Its 850 foot towers (the highest in Europe, and in a sensitive site, opposite Wren's Greenwich) are not subject to any planning control by local authorities. A new short take-off and land airport (London City Airport) is being built, and will open in 1987. The whole development climate has changed dramatically. In 1981, land was being offered at knock-down prices. Now riverside sites are selling for £7m a hectare.

The Government regards the LDDC as a tremendous success. It has, however, aroused violent opposition among the indigenous community. The ex-dockers argue that what they need is cheap rented council housing and new jobs which they could undertake. The new houses, they argue, are beyond their means, and are being sold to well-to-do people coming into the area. Most architectural commentators give verdicts on Docklands ranging from 'undistinguished' to 'a disaster'. The Corporation certainly provided the initial impetus for development, but development would eventually have come in any event. Land within sight of the towers of the 'City' (financial district) of London was unlikely to have remained vacant for long. The decision, in 1982, to build a light railway, linked with the Underground, into the docklands was a major factor in subsequent development; it will open in 1987. The Corporation has probably pursued development more vigorously than the local authorities would have done, although with less concern for the benefit to the locals. Finally, the way in which an unelected body has displaced elected councils must cause misgivings to those who still believe in local democracy.

The less well-known Merseyside Development Corporation was set up for a derelict part of the Liverpool docks. This was a much smaller area, in which no-one lived. It has consequently attracted less hostility, but it has also had less success. It has made the old docks a pleasant place, with leisure centres and shops, and a little housing, but it has attracted hardly any employment.

In April 1987, the Government announced that, in addition to taking over responsibility for town planning in their areas, the Urban Development Corporations would take over the larger council housing estates from the local authorities. In the words of one sympathetic newspaper, this proposal;

'brings together three important strands of Conservative policy. It continues the steady curtailment of the local authorities, advances the strategy of reducing public-sector housing to a minimum, and introduces a "business-friendly environment" to run-down inner-city areas. What is not clear is whether implementing the proposal would improve the management of publicly-owned houses for rent'. (Financial Times, 8 April 1987).

A somewhat less controversial model for the regeneration of depressed areas is a project also initiated by Mr Heseltine - 'Stockbridge Village' on the outskirts of Liverpool. This was

formerly a Liverpool overspill estate, 'Cantril Farm', consisting of tower blocks and 'maisonettes' built in the 1960s. By
1980, it was in a terminal state of vandalism, crime and
deterioration. With the assistance of substantial funding from
the Department of the Environment, the estate was taken over
by a public trust, which has undertaken physical alterations,
removing the more objectionable design features on the lines
indicated by Professor Alice Coleman. It has introduced a
more decentralised and effective system of maintenance, and
built some owner-occupied houses, but has retained a high
proportion of subsidised rented housing. It has not - as with
many 'privatisation' schemes - simply moved the tenants out
and sold off the houses to owner-occupiers, which 'solves' the
problem by moving it elsewhere. Although the scheme has its
critics, it has achieved an impressive change in the estate,
while retaining most of the tenants. However, no housing
improvements can change the fact that half the tenants are
unemployed.

The UDCs: A Critique
It has been argued that the UDCs - although at first sight at
the opposite extreme to the 'comprehensive redevelopment'
carried out by councils in the 1960s - in fact represent a
return to the ethos of that time (Wannop, 1985). Comprehensive redevelopment was based on purely physical improvement, and carried out 'from above' with little concern for the
local population. In the 1970s, there was a lot of talk (and
some action) about 'community programmes' and 'area depri-
vation initiatives'. Now an even more 'alien' organisation is
being imposed, with a remit primarily to encourage physical
development.

> 'It is a part both of the reversion to the physical and
> land development focus of urban regeneration projects,
> and of the growing participation of Government, that the
> benefits from the projects may now go increasingly not
> to people previously established in the area, but to
> outsiders. Of course, it could be said that this was
> common even twenty years ago when local communities
> were being displaced by CDA (Comprehensive Develop-
> ment Area) schemes, but when filling vacant land,
> rehabilitating and building new industrial buildings and
> private housebuilding bulks so large in current agency
> interventions, they connect much less clearly with local
> needs and provide much more opportunity for outside
> gain.'

This was written before the LDDC had got into its
stride; the hypothesis that the LDDC would, by its nature,
tend to benefit outsiders would seem to have been verified by

experience. It is highly probable that this experience will be repeated elsewhere.

TOWN PLANNING AND ECONOMIC DEVELOPMENT

Although the UDCs have been brought in because of the real or imagined defects of local authorities, the local authorities have by no means ignored the economic problems of inner areas. The sharp rise in unemployment, and its emergence as a political issue, has been reflected in local town planning policy. Various packages of assistance to industry have been made available by local authorities (within the limits of their straitened finances), and a new sub-profession of Economic Development Officers has emerged. Loans and grants, advance factories, advice to potential entrepreneurs, etc. are today available from many local authorities (Hausner, 1987). (The metropolitan authorities, abolished in 1986, were particularly active in this field.) In the Structure Plans for the large cities, employment creation has been placed high on the list of objectives, and decisions on planning applications frequently hinge on the employment associated with the proposal. Indeed, in many cities with high unemployment, almost any proposal which offers the prospect of jobs is likely to be accepted.

It must be added that this view of the town planning system is not shared by Government spokesmen, who criticise (Labour) local authorities for restricting employment through unsympathetic planning controls and holding on to land which could be sold to the private sector. Complaints of 'jobs locked in the filing cabinet' may have had some substance in the 1960s, but they are, we suggest, wide of the mark today (Williams and Gillard, 1980). Two recent White Papers, however, 'Lifting the Burden', and 'Building Businesses not Barriers' (HMSO, 1985, 1986) take a strongly anti-planning line. They propose both a dismantling of town planning and various positive measures - making land available for industrial development etc. - many of which are in fact already being implemented by local authorities.

The New Right v. Town Planning

The critique of local planning is in some cases based on a consistent intellectual concept, which rejects all constraints on the market system. Mr Nicholas Ridley, addressing British architects as Secretary of State for the Environment in 1986 said, somewhat Delphically, 'I hope the day will come soon when I can advocate freedom from all control', while Sir Keith Joseph, the politician who most influenced Mrs Thatcher, advocated, 'No controls except for sacred places'.

The issue, however, is not simply planning versus non-planning. Many people who reject laisser faire have criticised British planning, on several grounds; plans are overtaken by events; there is a lack of a clear framework of principles for individual decisions; planners are often ignorant of the practical problems of development. Proposals have periodically been made for planning 'from the bottom up' rather than 'from the top down' and for making planning conform more to 'the rule of law' by reducing discretionary powers. It was in response to these criticisms and proposals that Structure Plans were introduced in 1968. Since 1979, however, the planning system has been eroded rather than reformed. The central Government has inter alia been more ready to allow appeals by developers. The Structure Plan for Cardiff, for example, laid down the principle that 'hypermarkets' should be sited in regional shopping centres; 'free-standing' hypermarkets would not be allowed. But the developers whose proposals were rejected on these grounds appealed to the central Government, and all the appeals were allowed; the city is now ringed by hypermarkets. Similar stories can be told of other cities. At the moment, about half of all appeals are being allowed. One consequence of the Government's attitude is that developers now appeal as a matter of course.

Enterprise Zones

The Government has also created a number of 'enterprise zones'. These are more interesting for their background than for their results. In 1969 a group of urban scholars suggested - possibly tongue in cheek - that the debate on planning should be resolved by a controlled experiment; a large area, such as Yorkshire, should be freed from all planning controls, to see what happened (Banham et al., 1969). In the late 1970s, this suggestion was blended with other New Right ideas for abandoning health and safety regulations, minimum wages, company taxation etc. in limited areas, as a means of demonstrating the virtues of an 'enterprise economy'.

What finally emerged from this intellectual labour was a mouse. The 'enterprise zones' are areas of 100 ha. to 500 ha., in which planning controls are relaxed, and in which new enterprises pay no rates for ten years. The effect has been to cause enterprises to shift from outside to inside the zones, without bringing about much change in total employment in the cities in which they are situated. It is also clear from research studies that it is the financial incentives rather than the relaxation of planning controls which have attracted development (Tym, 1982-84; Lloyd, 1986).

The 1986 Housing and Planning Act contains provision for 'simplified planning zones' (SPZ) which offer the same relaxation of planning controls as EZs, but without the financial incentives. Although these 'zone planning regimes' originated from a strongly anti-planning ethos, this is not reflected in their practical implementation (Williams, 1986). General planning principles are laid down in a planning scheme for each zone, and the local authorities play a major role in land assembly and infrastructure development. The SPZs could - in the right political climate - become one means of ensuring as much planning as necessary, and as much freedom as possible.

Structure Plans
The latest Government proposal is to abolish structure plans, and replace them with a single tier of development plans drawn up by district authorities (DOE, 1986a). These plans would not require approval by the Secretary of State, but his regional officers would have the power to order changes in them. The counties will still be charged with preparing 'strategic policies' but will cede their control over local plans to Whitehall. Mr Ridley has justified the changes on the ground that 'The present two-tier system of structure plans and local plans is too cumbersome. Plans have become over-burdened with unnecessary detail and they take far too long to prepare'.

There is something in these criticisms, but many planners doubt whether the abolition of structure plans will 'provide a better basis for development control'. There will (critics maintain) be a loss of local democratic control, and new opportunities for confused responsibilities. A group of independent experts has recently published a (somewhat anodyne) report on the planning system (Nuffield, 1986). Its proposals on structure plans are superficially similar to those of the Government, but give a very different role to the central Government. Structure plans would be replaced by triennial 'county surveys', on which county 'development plans' would be based. District plans would be drawn up in conformity with the county plans. The role of the central Government would be confined to an annual White Paper on land use and the environment.

More radical proposals have been put forward by the British Property Federation, which represents commercial developers (British Property Federation, 1986). The Federation emphasises the need for allowing development except where it is clearly harmful; for ensuring that any refusals are based on legitimate planning, rather than on 'social engineering or blatantly political', grounds; and for providing clear rules so that developers know where they stand. It proposes that planning decisions should be limited by law to a consideration of 'relevant matters' - primarily whether the proposal conforms to the letter, or spirit, of the published

137

plans and guidelines. If there were appeals to the Secretary of State, he would be bound by the same principles. The Federation opposes the abolition of Structure Plans.

The Federation's proposals, although obviously 'business friendly', are not hostile to town planning as such. They are an interesting straw in the wind, in that they embody 'rule of law' concepts which are relatively new in British planning (although the Federation is arguably inconsistent in not proposing that appeals should be to the courts). Proposals on these lines could be one way of rescuing the British planning system from disintegration, but they would at present be unacceptable to both main political parties. Labour local authorities would reject as 'undemocratic' any legal limitation on their powers to control development, while Conservative national politicians would reject as 'undemocratic' any limitation on their power to overrule local authorities. For proposals of this type to be acceptable, there would have to be a reasonable measure of political consensus, and an acceptance, at least as an ideal, of the 'rule of law' in the continental European sense. As these conditions are largely absent, the proposals seem likely to remain 'academic'.

LAND PRICES

Complaints about 'ever rising land prices' have been common throughout the post-War period - and provided the motive force for the three abortive schemes for the taxation of land value. A long-term perspective provides a different perspective (Hallett, 1977). In real terms (as compared with either incomes or the cost of construction) prices fell sharply after the First World War, and in spite of catching up after the Second World War, were still below the pre-1914 level in the early 1970s. In the 'Barber boom' of 1972/73 there was a spectacular rise in nominal prices, and a doubling even in real terms (Figure 7.1). In the following two years, there was an equally spectacular fall. Since the trough of 1976, however, there has been a steady upward trend in real terms.

These average figures, however, can be misleading, because they conceal enormous regional differences - especially between prices in Greater London and the rest of the country. The average price in Outer London is four times the average for the rest of England and Wales, while the average price in Inner London is ten times as much (Table 7.1; Figure 7.2). These large differences in land prices reflect regional differences in house prices. Average house prices have traditionally fluctuated around 3.5 times average earnings; on this national basis, house prices were not unduly high in 1987. But a marked regional difference has emerged in the 1980s. Until 1982, the ratio was just over 4 in

London and the South East, and just over 3 in the rest of
Britain. Since 1982, the ratio in the rest of Britain (in
relation to national earnings) has remained constant, but the
ratio in London and the South East has risen to over 5.
Average household incomes, however, are now significantly
higher in the South East; if the figure is adjusted for London
earnings, it falls to around 4. This suggests that London
prices in general are not too 'abnormal'. They are, however,
double the level in many of the 'provinces', and the complaint
is increasingly heard that it is economically impossible to move
from the 'provinces' to London.

Why High Land Prices?
The 'mainstream' economic explanation of high land prices in
the South East is that high incomes increase the demand for
housing, and cause high house prices. The strong derived
demand for building land, combined with restrictions on
supply, then leads to high land prices. One land economist,
however, denies that high land prices in the South East
indicate a 'shortage' (Smyth, 1982, 1984a,b). Smyth makes
two points. Firstly, builders maintain private land banks,
which act as a 'buffer' between construction and demand. The
land price depends on how much land is being bought for
land banks, not on how many houses are being built. Smyth
appears to argue that the builders have artificially raised
prices in the South East through the manipulation of their
land banks. Secondly, he argues that the prevailing (1984)

Table 7.1: Prices of Residential Land (Bulk Land)

	Price, 1987 £ per ha.	Change in Price 1986/87 %
England and Wales (excl. London)	399,000	
Inner London	4,100,000	+51.2
Outer London	1,687,000	+44.5
South East	818,000	+28.1
East Anglia	625,000	+39.1
South West	440,000	+24.6
East Midlands	233,000	+26.2
West Midlands	245,000	+24.4
North West	192,000	+15.4
North	185,000	+12.3
Yorks and Humber	181,000	+13.1
Wales	137,000	+12.7

Source: Inland Revenue Valuation Office (1987)

Fig. 7.2: Average housing land prices for England and Wales
1970-85.

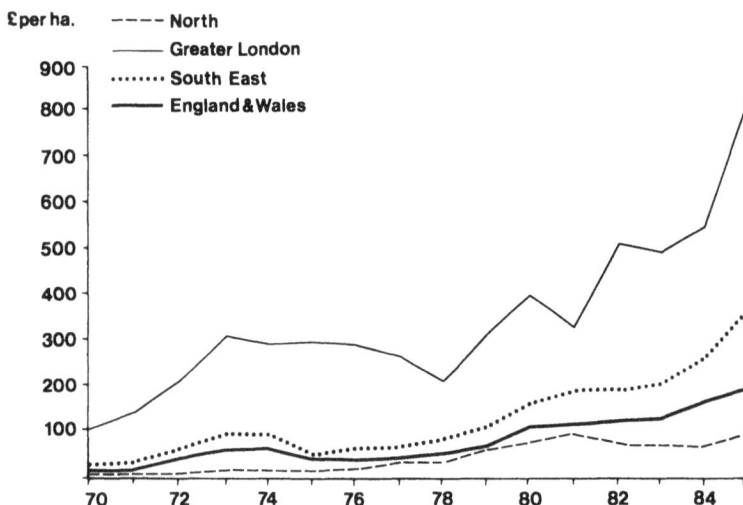

growth of GNP is unsustainable, and that housing demand is
likely to be correspondingly reduced.

It is certainly true that private land banks can introduce
'leads and lags' into the price of 'raw' building land. Moreover, if builders have bought too much expensive land, their
profits will be reduced; many firms went bankrupt after 1973
for this reason. If they have bought land cheaply, their
profits will be raised. But it seems far-fetched to argue that
the persistently high prices of land in the South East are due
to some manipulation by building firms rather than to the
pressure of demand. Regions of high income and growing
population, from Los Angeles to Munich, display high land
prices. More recently moreover, the (necessarily anecdotal)
evidence suggests that builders have reduced their land
banks to a minimum, and use 'options to buy' as a means of
securing future supplies.

Smyth is also right to stress that housing projections
should consider income as well as demography. However, his
prediction of a slackening of GNP growth has so far proved
incorrect. Real GNP, after falling 5 per cent in 1980-82, has
since risen at 2.5-3 per cent p.a. and - barring a marked
worsening of the world economic situation - seems likely to
continue growing at this rate, or a little less.

Some popular exponents of the 'long wave' theory have been even more apocalyptic than Smyth.

'Will the great Crash of 1986 catch you unawares - and your assets unprotected? ... What happens in the Down-wave? ... Share prices crash. ... Property prices dis-integrate, including YOUR house value'. (Beckman, 1983).

Well, it hasn't (yet) happened quite like that. In a few years, a slight downward trend in house prices in relation to income may well set in, as the population ages. It is only in the very high-priced districts, however, where site values account for 50 per cent or more of the price, that a 'collapse' seems at all likely. These districts include the 'fashionable' London districts (Kensington, Chelsea and, increasingly, Docklands) where prices rest to some extent on potentially fickle international demand. But this factor apparently does not play a predominant role.

'(A London estate agent) has 20 to 30 properties on his books at the moment that would cost you over £1m to buy, and in the single-million price range he finds that most buyers are now British people who can afford seven-figure homes because, like their country cousins, they have purposefully traded their way up-market, buying the best property they can afford each time, and selling and moving every couple of years'. (John Brennan, Financial Times, 2 May 1987).

THE SUPPLY OF BUILDING LAND

Underlying the controversies over land prices and develop-ment control is the recurrent argument as to whether the planning authorities have allocated enough land for housing. After a dialogue of the deaf between housebuilders and town planners on this question in the 1970s, a committee was set up representing both groups, which produced three reports (JLRC). The reports point out that, in 1984, approximately 11 per cent of the total area of England and Wales was in 'urban' use (including villages, roads etc.). At the present rate of growth, the figure will have risen to 12 per cent by the end of the century (JLRC, 1984). This extra 1 per cent would not seem necessarily to pose much of a threat to the countryside. (According to one poll, only 8 per cent of the population has an approximately correct knowledge of the level of land 'consumption'; 36 per cent put the figure more than ten times too high!) There is, however, a problem, in that the most rapid growth is in the South East, where the problems of congestion, and the constraints on growth, are greatest.

In 1983, the Committee concluded that England and Wales should plan for 215-220,000 dwellings per annum, till the mid-1980s, and probably beyond. This was slightly below actual starts, but less than the estimates of the 'necessary' number of starts given in the 1977 'Green Paper' (DOE, 1977c), the latest comprehensive review of the housing situation. The housebuilders' representatives on the Joint Committee maintained that the Structure Plans provided for only 180-190,000 dwellings, and so were too restrictive, whereas the planning officers maintained that they allowed for the target number. These aggregative studies of 'need' and land availability are clearly subject to a large margin of error. Moreover, they take no account of the composition of housing starts, in particular the steady elimination of council building. An examination which ignores sub-markets avoids some of the most acute problems; the fact that there is an ample supply of 'executive homes' has little relevance to a family on Supplementary Benefit.

One point on which there was agreement was that, although inner city building and renovation could make a valuable contribution, it would not be sufficient by itself. The annual contribution of inner areas was unlikely to exceed 3 per cent of housing output, while output within built-up areas was unlikely to exceed 15-20 per cent.

> 'Most of the remaining 80 per cent of production will continue to be on sites on the periphery of cities and larger towns, and with and around smaller towns and villages. Depending on how one uses the English language, some of these sites may or may not be described as "greenfield". Most will not, however, be deep in rural countryside. With wise use of the planning system, none need be in green belt.' (JLRC, 1984, p. 7).

It is clear that the 'macroeconomic' assessment of housing demand and land availability is problematical. There is, however, an alternative (or complementary) 'microeconomic' approach. When the cost of land - as a percentage of building costs - rises to 'abnormal' levels there is a prime facie case for zoning more land for housebuilding. The Committee points out that the 'traditional' percentage has been around 15 per cent, and recommends that, when a higher figure prevails, the planning authority should consider whether more land should be allocated. There is clearly no magic about 15 per cent, but if the typical percentage is 30 per cent or more, there is a case for considering a change of policy. On this basis, there would seem to be a case for allocating more land for housebuilding in the South East, but not in other regions.

The Housebuilders' Inner City Inquiry

In spite of the evidence that inner-city land could make only a partial contribution to total needs, there have been continued demands that more new housing should be sited in inner-cities. As a result, the House Builders' Federation set up an inquiry in 1986 to examine the scope for new housebuilding in inner areas. Its preliminary conclusions stressed the fundamental difference between London and other cities, especially those in the North of England. In the words of Mr Roger Humber, the HBF director, 'The key problem in London is not making the sites economically worthwhile, but a political problem, as the local authorities are deeply distrustful of housebuilders, and refused even to cooperate with the inquiry'.

However, there is some ground for the suspicions of the (Labour) London councils - quite apart from the fundamentalist objection to all private house-building to which some of their members undoubtedly subscribe. Alongside a seemingly endless supply of people who can afford low six-figure prices, London has the largest concentration of homeless people in the country, and districts in which the unemployment rate is 35 to 40 per cent. Many people cannot afford to buy a dwelling at current London prices. On grounds of equity, there is a case for using a substantial proportion of the available land for subsidised rented housing, rather than using it all for building houses for sale, as is tending to happen under present Government policy.

The situation in other major cities is very different. In Mr Humber's words, 'The overwhelming problem outside London - particularly in the North with its high unemployment - is the inability of residents in inner urban areas to afford the cost of a new house when old houses are so much cheaper'. For those who can afford a new house, the inner areas are not attractive. 'Who wants to buy a new house in the middle of Middlesbrough surrounded by chemical dumps and waste storage when you are only ten minutes' drive from the villages in the Cleveland hills?'

The final report concluded that private housebuilding in inner cities would not take on the scale desired by the Government unless the amount of public investment were substantially increased; it recommended a doubling of the funds available for housing programmes (House Builders' Federation, 1987). It also recommended the setting up of 'land agencies' with powers to purchase land compulsorily, provide services, and reclaim land for private housing, in areas where local authorities were unwilling or unable, for financial or political reasons, to undertake the task. The proposed agencies seem similar to the 1965 Land Commission (sic), but would be organised on a regional basis, e.g. for London, or for the North of England (see also Chisholm and Kivell, 1987).

NORTH AND SOUTH

A theme which runs through all reports on the land market is the difference between economic conditions in 'the North' and 'the South' - the division being roughly a line from the Bristol Channel to the Wash. (The division is not clear-cut. There are pockets of affluence in the North and pockets of poverty in the South, but the average statistics show a significant divergence.) In the North, the entire economic base of many cities has been lost; although publicly-financed environmental and housing improvements are necessary, they will not solve the problems unless unemployment is cut dramatically.

One proposed solution to Northern unemployment is migration to the South, but there are limits to the extent to which this is feasible. Regional differences in housing costs are so great that North-South migration is often worthwhile only if a well-paid job can be guaranteed, and that is far from the case. There is substantial unemployment in the South - 400,000 in London alone. A slight migration is nevertheless taking place from the Northern regions to the Southern regions - outside the South East itself. A far bigger migration is taking place from London to other parts of the South-East region, and to adjoining regions, such as the South West and East Anglia. This migration to 'near London' has implications for housing and land supply.

There are areas in 'the South' where, to judge by land prices, more housing should be supplied, but a building 'free-for-all' could recreate the problems of the 1930s on a much larger scale. In the 1960s and 1970s, there was a consensus among planners and geographers that the way to resolve this dilemma was to concentrate development along transport lines, in planned development which limited the 'damage'. Proposals were also made for making 'green belts' more worthy of the name. Merely preventing development did not always produce attractive countryside. It would be better, some people argued, for public bodies to acquire rural areas near towns and manage them with a view to amenity. These ideas produced relatively few results, but some steps were taken to increase the amount of 'amenity land' by bodies like the Forestry Commission, and the National Trust (a private charity).

At the present time, a modest programme of 'new and expanded towns' could cope with the demand for housing in the South in an environmentally acceptable way, which would be a complement rather than a substitute for urban renewal (and job creation) in the North. At the same time, publicly owned areas of forest (as in West Germany) or heath, or farmland, could be maintained around towns, managed with a view to amenity. (The vast majority of rural land would remain in private ownership, and even the publicly owned land could in many cases be let to tenants, under appropriate tenancy agreements.)

There is little immediate prospect of policies of this sort being adopted for recreational areas. The Government has forced local authorities to sell off playing fields and small-holdings, and the Forestry Commission to sell some woodland. The emphasis is wholly on privatisation. On the other hand, there are some moves toward a new breed of 'new towns'. The Government, in spite of its commitment to deregulation, is under strong political pressure to preserve the London 'Green Belt'. Thus the Secretary of State for the Environment refused permission, in 1986, for a privately developed 'country town' of 14,000 inhabitants in the London Green Belt at Tillingham Hall. But there are areas further from London where 'country towns' would be acceptable to the Government, and defensible on planning grounds. Consortium Developments, a group formed by the nine largest housebuilders, has launched a campaign to get planning permission for a 'country town' of 11,500 people at Foxley Wood, on an old gravel pit in Hampshire. Another consortium is already building what is described as 'Europe's largest private new town' - a community of 25,000 people, called Bradley Stoke, north of Bristol.

There is a danger that these private 'country towns' will not have the social mix and the variety of employment that the original New Towns at least aspired to, and will simply be a new form of dormitory suburb. But if they were well planned, they might well make a useful contribution to the supply of housing in the South, in an environmentally acceptable way. It certainly seems worth carrying out some experiments. The essential proviso is that an effective planning system is retained. In some cases, public investment is the pre-requisite for private investment. The Cardiff Bay project, for example, is based on a small barrage, which would provide the basis for a marina and new housing. But even though the Cardiff Bay project (organised by a UDC) is certainly worthwhile, it is unlikely to be of much benefit to the inhabitants of the desolate council estate which replaced 'Tiger Bay'.

'Concreting over the Countryside'
The Government has also had to undertake a re-examination of 'urban' development in rural areas. Till now, planning guidelines have been based on keeping the countryside for agriculture, and restricting housing and non-agricultural development. On the other hand, agricultural land use is not subject to any planning controls. With the decline in the agricultural labour force, and the collapse of many rural crafts, this 'agriculture first' policy has been increasingly criticised. Does it make sense, it has been asked, to refuse

to allow redundant barns to be used for housing or work-shops? Moreover, with the growing surpluses of agricultural products, substantial areas of agricultural land could be freed for other uses - forest, conservation, recreation, etc.; one study concludes that up to 23 per cent of agricultural land could be available for other uses by the year 2000 (Country-side Commission, 1987).

In 1987 the Government began considering a secret report code-named ALURE (Alternative Land Use and Rural Economy), and there were claims that this report proposed abandoning all planning controls in the countryside. When the Government's proposals were announced, they proved to be very modest - although they produced strong criticism. The Chairman of the National Trust claimed that 'two-thirds of the countryside is now at risk from the slow, invasive crawl of urban development'. The new guidelines allow housing and other types of development to be considered, in place of the previous exclusive concern with agricultural production [DOE, 1987). The consequences seem likely to be undramatic, but potentially beneficial. A few more workshops should help to alleviate the lack of employment, while sensitive small-scale housing developments would meet a demand, without necess-arily disrupting the environment. In short, the change of emphasis should be beneficial - provided that a planning framework remains.

THE COMMUNITY CHARGE

The latest step in the 'radical Conservative' programme is the abolition of local 'rates' (property tax), and the substitution of a poll tax, or 'Community Charge', which would be the same for all adults (DOE, 1986b). The rating system has long been criticised, both because any property tax arouses host-ility when too much weight is placed on it, and for technical shortcomings peculiar to Britain. The criterion for 'rateable value' - laid down in the 19th century - is the annual value at which the dwelling could be let. Since private dwellings are now rarely let, this criterion has become meaningless; Rating Officers appear to use floor space and subjective assessments of capital value. The obvious solution - the formal adoption of market value (i.e. sale price), as in the USA - has been proposed for decades. But although the existing method of assessment leads to some anomolies, it does rough justice; larger and more valuable houses generally have higher rateable values than smaller and less valuable houses. The average annual payment is around 1 per cent of capital value.

Rates have the virtue of being cheap to collect, and (as they are paid by the occupier) virtually evasion-proof. It has long been clear, however, that they are an inadequate sole basis for local finance, and additional, or alternative, taxes have been recommended for consideration (DOE, 1976). A poll tax has never been recommended; indeed, in 1983 the Government itself rejected it as unworkable (DOE, 1981b, 1983). The main criticism of the poll tax is that it is regressive. The payment is likely to be around £250 a head per year. A single person with an income of £20,000 a year will pay 1.25 per cent of his income, but a pensioner couple with an income of £4,000 will pay 12.5 per cent. Even if rebates were introduced for the very poor, the benefits for the (house-) rich would remain. There will either be a marked regressive change in the tax system or, as the Association of District Councils argues,

> 'the community charge will be so rapidly discredited ... as to necessitate another expensive reform of local taxation within a short period of time'.

In addition to redistributing income, the substitution of a 'community charge' for 'rates' involves the abolition of an indirect tax on housing. The consequences of such a change are - as any textbook of micro-economics will show - likely to be a rise in price and/or an increase in supply. Since, however, the 'tax reduction' is greatest for the most expensive houses, the increase in price is likely to be concentrated in this sector. One study estimates that average house prices in various regions will rise by 12 to 23 per cent (Hughes, 1987). It seems a curious change at a time when there is considerable concern about house price inflation.

A LAND REGISTER?

In spite of the vast amount of legislation on urban land which has been passed and repealed since 1945, there are certain basic public institutions which have not been created, or developed adequately, because of governmental neglect. One is a public Land Register. In Scotland (which has a different legal system), there has been a Register for centuries. In England and Wales, on the other hand, the setting-up of a Register was long opposed by landowners and lawyers. There is now a Land Register of sorts, and property is (eventually) registered when it is sold. (There is a serious back-log because of Government restrictions on staffing, even though the registry makes a profit.) However, the information is limited to the physical details of the site and the name of the owner; more importantly, this information can be released only with the permission of the owner. It is thus impossible (or at least very difficult) to obtain any information on the pattern of ownership. A proposal has been made by the Law Commission to allow public access to the Register (Law

Commission, 1983). This proposal is opposed by the Country Landowners' Association ('an invasion of privacy') and the Law Society. The British Property Federation supports public access to the Register, except for financial details.

Note added in proof
Legislation in 1988 gave public access to the register.

CONCLUSIONS

Since 1945, Britain has experienced a continual see-saw in land and housing policy. The lack of continuity is most plausibly explained by the country's unique system of government. As a German academic points out, in another context;

> 'British government can indeed be characterised as being highly concentrated and centralised with a pattern of decision-making that can be characterised as "unilateral" and highly "exclusive", allowing the almost complete disregard of the positions presented, and the interests represented, by the opposition parties' (Scharpf, 1981).

The policies pursued since 1979 have combined privatisation and deregulation with administrative centralisation on a scale never before seen in Britain in peace-time. Very different assessments have been made of 'the Thatcher experiment'. Supporters argue that it has produced an 'economic renaissance' (Walters, 1986); that the Government has protected the people from the waste and tyranny of Labour local authorities; that the spread of owner-occupancy has been socially desirable, and the greater freedom for developers economically beneficial.

On the other hand, critics argue that Disraeli's 'two nations' have been re-created; that the infrastructure has been neglected; and that sound town planning has been prejudiced. It has been clear for some time that the traditional role of town planning needed to be re-examined, but planners have on the whole adapted quite well to changing conditions, by assuming a more positive and competitive role in negotiating development and implementing projects. The Government, however, believe that planners and local authorities are 'obstacles'. As Mr David Trippier, the Minister for the Inner Cities, has put it (at a conference held, appropriately enough, in London Docklands),

> 'New and growing businesses have avoided areas of high rates, drab public sector housing, poor schools and in far too many cases incompetent and downright loony local governments. ... We wish to sweep away the obstacles to development ...' (30 June 1987).

As part of this process, miniature Urban Development Corporations were to be created.

If all goes according to the Government's (now quite explicit) plan, within five years local government - deprived of most of its responsibility for public housing, education, town planning, and public transport, and with many other functions compulsorily privatised - will be a shadow of its former self; the management of the rump of public housing, and many public land management matters, will be in the hands of 'business friendly' central government nominees; and the benefits of this state of affairs will be generally recognised. On the other hand, given Britain's past, it would be rash to predict its future. It would be one of history's not unfamiliar jokes if the apparatus of centralised control built up since 1979 were eventually to be inherited by the Labour Party. It seems safe to say, however, that there is little prospect in the foreseeable future of the adoption of consensual, 'mixed economy' land-use policies, such as are found in the Netherlands, Germany or France.

REFERENCES AND FURTHER READING

Aldridge, M. (1979) The British New Towns, London
Ball, M. (1985) Housing Policy and Economic Power, London
Banham, R. et al. (1969) 'Non-Plan: An Experiment in Freedom', New Society, 20 March 1969
Beckman, R. (1983) The Downwave, London
Boleat, M. (1986) Housing in Britain, Building Societies Assoc., London
British Property Federation (1986) The Planning System - A Fresh Approach, London
Burgess, T. and Travers, T. (1980) Ten Billion Pounds; Whitehall's Takeover of the Town Halls, London
Burns, W. (1983) 'The Planning System; Are Major Changes Needed?', National House-Building Council, Conference on Housing and Planning, London
Coleman, A. (1985) Utopia on Trial; Vision and Reality in Planned Housing, London
Child Poverty Action Group (1987) The Growing Divide - A Social Audit 1979-1987, London
Chisholm, M. and Kivell, P. (1987) Inner City Waste Land, Hobart Paper 108, IEA, London
Countryside Commission (1987) New Opportunities for the Countryside, Manchester
Crawford, P., Fothergill, S. and Monk, S. (1985) The Effect of Business Rates on the Location of Employment, University of Cambridge
Cullingworth, J.B. (1981) Peacetime History: Environmental Planning 1939-69, Vol. IV, Land Values, Compensation and Betterment, London, HMSO

149

Cullingworth, J.B. (1985) Town and Country Planning in England and Wales, London
DOE (1965) (Department of the Environment), The Land Commission, Cmnd 2771, HMSO, London
DOE (1974) Land, Cmnd 5730, HMSO, London
DOE (1976) Report of Committee of Inquiry into Local Government Finance (Layfield), Cmnd 6453, HMSO, London
DOE (1977a) Policy for the Inner Cities, Cmnd 6845, HMSO, London
DOE (1977b) Inner Area Studies: Liverpool, Birmingham and Lambeth; Summaries of Consultants' Final Reports, HMSO, London
DOE (1977c) Housing Policy; A Consultative Document, Cmnd 6851, HMSO, London
DOE (1981a) Planning Gain - Report of the Property Advisory Group, HMSO, London
DOE (1981b) Alternatives to Domestic Rates, HMSO, London
DOE (1983) Rates, HMSO, London
DOE (1986a) The Future of Development Plans, HMSO, London
DOE (1986b), Paying for Local Government, HMSO, London
DOE (1987), Farming and Rural Enterprise, HMSO, London
Douglas, R. (1976) Land, People and Politics, London
Economist (1982) Britain's Urban Breakdown, London
Edinburgh, Duke of (1985) (Chairman) Inquiry into British Housing, London
Evans, H. (1972) New Towns; the British Experience, London
Eversley, D. (1974) 'Britain and Germany; Local Government in Perspective', in R. Rose (ed.), The Management of Urban Change in Britain and Germany, London
Goodchild, R.N. (1985) Development and the Landowner, London
Hallett, G. (1977) Housing and Land Policies in West Germany and Britain, London
Hallett, G. (1979) Urban Land Policies; Principles and Policy, London, Chap. 9
Hausner, V. (1987) Urban Economic Development and the Future of British Cities, Oxford
HMSO (1985) Lifting the Burden, London
HMSO (1986) Building Businesses not Barriers, London
House Builders' Federation (1987) Private House Building in the Inner Cities, London
Howard, E. (1898) Tomorrow, published as Garden Cities of Tomorrow, Osborn (ed.), London 1945
Hughes, G. (1987) 'Rates Reform and the Housing Market', Edinburgh University Discussion Paper
Inland Revenue Valuation Office (1987) Property Market Report, RICS Publications, London
Jones, G. and Stewart, S. (1985) The Case for Local Government, London
JLRC (1982) (Joint Land Requirements Committee), Is there Sufficient Housing Land for the 1980s? Paper 1. How

Many Houses Should We Plan for? Housing Research Foundation, London

JLRC (1983) Is There Sufficient Housing Land for the 1980s? Paper 2. How Many Houses Have We Planned for? Is There a Problem? Housing Research Foundation, London

JLRC (1984) Housing and Land; 1984-1991;1992-2000. Housing Research Foundation, London

Keogh, G. (1985) 'The Economics of Planning Gain', in S. Barrett and P. Healey (eds), Land Policy: Problems and Alternatives, Oxford

Law Commission (1983) Property Law; Land Registration, HMSO

Liberal Party (1925) Towns and the Land, London

Lloyd, M.G. (1986) 'The Continuing Progress of the Enterprise Zone Experiment', Planning Outlook, 29(1)

Loughlin, M. et al. (1985a) Half a Century of Municipal Decline 1935-85, London

Loughlin, M. (1985b) 'Apportioning the Infrastructure Costs of Urban Land Development', in S. Barrett and P. Healey (eds), Land Policy: Problems and Alternatives, Oxford

Manser, W.A.P. (1985) The British Economic Base, London

Mayes, D. (1979) The Property Boom, Oxford

Montgomery, J. (1987) 'The Significance of Public Landownership', Land Use Policy, London, Jan

Moor, N. (1985) 'Inner City Areas and the Private Sector', in S. Barrett and P. Healey (eds), Land Policy: Problems and Alternatives, Oxford

Newton, K. (1981) 'The Local Financial Crisis in Britain', in L.S. Sharpe (ed.) The Local Fiscal Crisis in Western Europe, London

Norton, A. (1983) The Government and Administration of Metropolitan Areas in Western Democracies, University of Birmingham, UK

Nuffield Foundation (1986) Town and Country Planning, London

Prest, A.R. (1981) The Taxation of Urban Land, Manchester

RICS (Royal Institution of Chartered Surveyors) (1974) The Land Problem - A Fresh Approach, London

RICS (1978) The Land Problem Reviewed, London

Riddell, P. (1985) The Thatcher Government, London

Scharpf, F.W. (1981) The Political Economy of Inflation and Unemployment in Western Europe: An Outline, Wissenschaftszentrum, Berlin

Smyth, H.J. (1982) Land Banking, Land Availability and Planning for Private Housebuilding, SAUS Working Paper 23, University of Bristol

Smyth, H.J. (1984a), Land Supply, Housebuilders and Government Policies, SAUS Working Paper 43, University of Bristol

Smyth, H.J. (1984b), Property Companies and the Construction Industry in Britain, Cambridge University Press

Travers, T. (1987) The Politics of Local Government Finance, London

Tym, R. and Partners (1982, 1983, 1984) Monitoring Enterprise Zones, London

Walters, A. (1986) Britain's Economic Renaissance, Oxford

Wannop, U. (1985) 'Government Agencies in Land Development', in S. Barrett and P. Healey, (eds), Land Policy: Problems and Alternatives, Oxford

Williams, R.H. (1986) 'Zone Planning Regimes', Planning Outlook 29(2)

Williams, R.H. (1987) 'Tyne and Wear's Industrial Improvement Areas', Northern Economic Review 1(4), Newcastle

Williams, R.H. and Gillard, A. (1980) 'Employment Planning; Powers and Policies in Tyne and Wear', Planning Outlook, 23(2)

Chapter Eight

LAND POLICY IN THE UNITED STATES

David E. Dowall

INTRODUCTION

Writing about the land policy of the United States is a trifle presumptuous. There is no de jure land policy. In fact, there have been only two periods in US history when the Federal government set out a clear land policy: the 1850s and 1860s (the Homestead Act of 1862), when the vast Western public lands were settled, by permitting settlers to claim ownership to land after occupying and using it for a period of time; and the 1920s, when the Federal government facilitated the adoption of land-use 'zoning' by local governments (deNeufville, 1981; Wolf, 1981).

On a more implicit level, the Federal government engaged in national planning during the Depression, developing Greenbelt towns and agricultural villages (for example through the National Resources Planning Board). In the late 1960s and early 1970s, concern over the failure of urban renewal, environmental degradation, and a seemingly imminent energy crisis fostered two serious attempts by the Federal government to promulgate national land-use policy - one by the Senate in 1976 during the Ford administration and one later by the Carter administration, both of which were short-lived (Hagman, 1980; Popper, 1981). Many local communities, however, imposed 'growth controls' on housing development. The Federal government has generally stayed out of the picture, leaving such matters to States, cities and counties. This reflects the separation of powers in a federal system and, more importantly, the long-term commitment to the marketplace as a means of allocating land resources.

In the 1980s the nation's attention has shifted to economic development, job creation and community revitalisation. Now many land-use planners are interested in promoting office centres, industrial parks and housing projects, not in using growth management to slow down development. Perhaps the most interesting current trend in land-use policy is the diversity of policy responses. In some depressed

153

central cities, expansive land development, including land banking, land write-downs (i.e. sales of land at below acquisition cost) and tax holidays are common. In others, stringent land-use regulations, such as growth controls, low-density limitations, and extensive open space requirements, are the norm. This diversity of land policy is the result of the growing disparity of urban development across the nation, and its differential impacts on communities. These differences are not purely sunbelt-frostbelt in character, as the growth of Boston and Philadelphia, and the decline of Houston, attest. In some cases, boom-and-bust examples abound within one metropolitan area, leading to conflicting or contradictory policies, given the absence of local government coordination.

In some instances, local land-use policy has inhibited community development; in others, a lack of coordination has caused growth problems - especially traffic congestion and lack of affordable housing. In fact, one of the major problems facing high-growth metropolitan areas is the unbalanced distribution of employment centres and residential areas. Some areas have large and growing job bases and little or no housing, while others are exclusively residential. Such patterns cause enormous traffic congestion problems, as people are forced to commute great distances, and as the scattered patterns of employment and residence make it much more difficult to design and implement effective transportation systems.

What is becoming quite clear to many US land-use planners is that planning tools effective in one town are no longer automatically effective in others. Planners across the nation are searching, on the one hand, for new ways to link land policy to economic development and, on the other, for more effective methods to regulate urban and suburban development (Committee on National Urban Policy, 1983).

This chapter provides a brief overview of land-use controls and land-use policy issues in growing and declining America. It begins by outlining the institutional context of land-use planning, covering Federal, State and local land-use policies. It then proceeds to outline the structural changes which are reshaping US cities and regions, their land-use impacts, and the policy implications.

PUTTING THE US SYSTEM OF LAND-USE CONTROLS INTO CONTEXT

In the United States, the development of land is determined by market forces and the regulatory power of a myriad Federal, State and local government entities. The Federal Government once owned huge areas in 'the West'. which have since been disposed of, and it still owns nearly one third of the land area of the USA, mostly forest or desert land; some of this land consists of National Parks. Most land in the neighbourhood of cities, however, is privately owned, and most land transactions take place between individuals and private corporations. US cities have not followed the Germanic and Scandinavian approach of acquiring 'land banks'. The subject was widely discussed in the 1970s, but little action was taken.

There is a long-established right of 'eminent domain' (meaning 'highest authority') under which the Federal Government, or State Governments, may acquire land compulsorily (or 'condemn' it). There are differences in the laws of the various States but all provide for a right of 'eminent domain', provided that the land is to be used for a public purpose, and that fair compensation is paid. The right may be delegated to public agencies, local authorities, or even a private corporation. In general, the owner of land must be invited to sell it, and refuse the offer, before the power of eminent domain can be invoked. All condemnation proceedings call for court action. Compensation is decided by the courts, usually on the basis of 'fair market value'. Since the 1940s, it has been accepted in most States that land may be acquired for a purpose of public interest, rather than merely for 'public use', provided that the land is subsequently disposed of. In practice, however, 'eminent domain' is mainly used in connection with highways and similar 'public works'. It has on occasion been used to acquire old housing for urban renewal, but it is hardly ever used to acquire 'greenfield' sites for housing development.

Zoning

Although the Federal Government has made little attempt at promulgating a national land policy it did, in the late 1920s, assist in the development of the principal form of land-use control, known as 'zoning'. In 1928, the US Department of Commerce drafted a model ordinance of State legislation, permitting local zoning controls over land-use (National Commission on Urban Problems, 1968). By the 1930s, most states had adopted a version of this model ordinance, and local planning commissions became the principal regulators of land-use. The Federal Government should not be credited with 'giving birth' to US zoning, since many communities were zoning before the 1928 ordinance was drafted. It did, however, clearly facilitate the spread of zoning.

'Zoning' means that an area in which new development takes place can be divided into 'zones' specifying the type of development permitted, e.g. 'detached single family residences', 'multi-family residences', 'industrial development', 'commercial development' and various sub-categories. It is essentially a means of giving some order to development (and keeping undesired uses out of prime residential neighbourhoods); it does not involve any power to prohibit development as such. In most rural communities, there is virtually no means of preventing 'strip development', 'ranchlets' and other forms of 'scatteration'. Thus 'zoning' is a weak tool for protecting specified rural areas from development, or for concentrating development in 'balanced' communities, with provision for employment and recreation, and a range of types of housing. Zoning also has little relevance to the regeneration of decayed older urban areas.

Federal Policies

There have been several recent attempts to address urban and environmental problems, and these initiatives contain much on land policy. The Advisory Commission on Inter-governmental Relations, in a study released in 1968, triggered extensive discussions on the need for national growth policies (ACIR, 1968). The Commission observed that the nation's development patterns were the result of countless public and private land-use decisions, but that the Federal Government, by funding community infrastructure, deciding where to locate governmental and defence installations, and allocating pro-curement contracts, exerted great influence over patterns of land utilisation.

The debate over the appropriate federal role spawned the Urban Growth and New Communities Development Act of 1970. This act mandated the Executive Branch of the Federal Government to examine patterns of land-use and urban growth, publish a biennial report from the President to Congress analysing trends and problems, and recommend how the government and the private sector should respond to such problems.

The President's report on National Urban Policy was submitted to Congress in 1972, 1974 and 1976. During the Nixon administration (1968-74), the main thrust was to abolish categorical grants, set up six block grant programmes, and reorganise the Cabinet, creating four new 'super agencies' to handle domestic issues, one of which, the Department of Community Development, was to combine the functions and responsibilities of housing, transportation, commerce and agriculture. The plan threatened so many entrenched interests that it was not taken seriously, and the Watergate scandal finally killed it.

The Ford Administration (1974-76) adopted a traditional posture towards land and urban policy, advocating a national commitment to the preservation and restoration of central cities. It continued the Nixon Administration's program of replacing federal categorical grant programs with block grant programs.

It was with the Carter Administration that a new interest in urban policies emerged. The Carter Administration attempted to develop fewer but more comprehensive programs for urban revitalisation. Policy 10 of the 1978 biennial report is sharply focused on land-use matters; 'Federal laws and programs will be amended to discourage sprawl' (The President's National Urban Policy Report, 1978). This was to be accomplished in three ways: by coordinating federal and state actions affecting land-use development, stimulating metropolitan-wide planning and development programs, and offering strong incentives for businesses to remain in central city areas.

The cornerstone of the Carter urban program was the urban impact assessment (UIA) (Glickman, 1980). This was to be conducted for any policy, program or funding award which might have deleterious effects on cities. One example was the use of Community Development Block Grant funds for the financing of infrastructure for suburban shopping centres, which allegedly siphoned retail and commercial activities from central cities to suburbs. Studies of at least a dozen shopping centre projects were made and, in once instance, federal funds were blocked (Teitz and Dowall, 1980).

This is not to say that UIAs were a failure, but rather that it is difficult to use a case-by-case approach to set national land-use policy goals. There were far too few opportunities for using the policy to shape patterns of development; the UIA was much less effective than the 'environmental impact statement' (EIS), used by the Federal and most State governments to evaluate projects. Thus, despite several attempts at adopting a national land policy, the nation has been left with little more than the 'zoning' system adopted in the 1920s.

With the Reagan Administration (1980), greater emphasis was placed on limiting the role of government in society. Accordingly, many domestic programs were eliminated or scaled down; the Carter 'Urban Policy Agenda' was scrapped. The Reagan agenda is designed to shift governmental programs to States and local governments, and this coincides with earlier efforts of environmentalists to establish strong land-use policies at the State level.

It should be mentioned that the vast majority of housing is in the private sector. Nearly two thirds of American households own their homes, which are mainly detached wood-framed 'houses'. In the more affluent areas there are also modern, privately rented apartment blocks, which cater mainly for singles. There is, however, a fiscal incentive to home ownership, in that mortgage interest payments (like all interest payments) are tax-deductible, while there is no tax on 'imputed rent'. In the run-down central city areas, the old apartment blocks are privately owned, and often neglected. The basic problem of these areas, however, is the concentration of poverty, unemployment and social breakdown. The Federal presence in the public housing area is declining, and public housing has become 'last resort' shelter; the median income of public housing tenants stands at 28 per cent of the national median income. Thus, as far as national housing policy goes, the US has decided to get out of the business of providing housing for those who need it but cannot afford it.

The Quiet Revolution in Land-use Controls

In the early 1970s, America woke up to the fact that it was rapidly destroying its stock of physical resources. Air and water pollution had become acute, and when the Federal Government adopted national standards for air quality, many metropolitan regions discovered that they failed to meet them. Erosion carried away sediment and raw sewage, while industrial waste poured into lakes and streams, killing fish and waterfowl. At one point, the Cuyahoga River in Cleveland become so polluted that it caught fire. Rapid suburbanisation had severe impacts on small towns and villages on the edges of metropolitan areas, causing traffic congestion and threatening the small-town quality of life that had attracted many in the first place.

Around 1970, public sentiment shifted, as citizens demanded stricter environmental controls at all levels of government (Reilly, 1973). The US Environmental Protection Agency stepped into action, promulgating stringent standards for pollution control. For many, the passage of the National Environmental Policy Act of 1969 marked a threshold in federal action. But the most important aspect of the environmental movement was the establishment of a State (as distinct from Federal) presence in land-use and environmental controls. As early as 1966, prominent planners were calling for the establishment of State agencies to control land development (Popper, 1981) and, later on, several public and quasi-public bodies echoed this call (National Commission on Urban Problems [1968], Rockefeller Brothers Fund Task Force [Reilly, 1973], and Ralph Nader's Study Group on Land Use in California [Fellmeth, 1973]). The two main professional bodies concerned with land-use issues, the American Bar Association and the American Planning Association, endorsed State-level land-use planning and policy-making. The American Bar Association went one step further: it developed a Model Land Development Code (American Law Institute, 1976). Governments in Florida, Hawaii, Colorado, California and Vermont enacted State planning controls.

By 1975, 27 of the 50 States had adopted new programs involving either State land-use planning or State review of local land-use decision-making. Most of the State-level land-use attention has focused on rural areas; few State-level efforts have had any effect on urban land-use issues in metropolitan areas. Leaders in State-level land-use planning -

Hawaii, Vermont, Florida, California, Oregon and North Carolina - enacted land-use controls because of second home development or expanding urban development onto agricultural lands (DeGrove, 1985).

Local Growth Controls

A parallel development to the spread of State-level land-use regulation was the proliferation of growth management programs in cities across the US. The interest in growth management is understandable, given the inability of zoning and general plans to control or modulate the temporal rate of development.

Most States require cities and counties to prepare and update community master or general plans (Kent, 1964). These plans outline anticipated population and economic growth, and designate which areas of the community are appropriate for development. These designations reflect not only potential growth and the resulting demand for land, but also the environmental capabilities of the land to support development. In most communities, zoning is the principal method of implementing the general plan. The zoning ordinance sets the permissible use and density of development, and development permission is normally denied if the proposed project is not consistent with the zoning. Thus zoning and general or master plans control the type and density, but not the rate at which development may occur.

In some communities the pace of development has been torrid, and citizens have pressed their planning commissions to limit it. Pinpointing exactly where this movement started is fairly easy. The cities of Ramapo, New York and Petaluma, California, in 1972 and 1974 respectively, questioned the prevailing wisdom of unbridled growth, and unilaterally imposed limits on the rate and level of residential development (Scott, 1975). After this bellwether event, numerous communities across the nation followed suit. At the height of the local growth control movement, over 200 cities and counties claimed to be actively limiting development (Dowall, 1980).

This early version of growth control focused exclusively on residential development, not commercial or industrial development. In virtually every case, the growth controls were imposed by suburban towns in close proximity to major central cities. There is a very good reason for these patterns. The suburban growth control community is usually inundated with residential development, but it does not need an economic base in order to flourish. It can 'borrow city size' by relying on the nearby central city for employment and business services (Alonso, 1973). The spread of local land-use controls was rapid, and in some regions the suburbs became bastions of growth control. Coupled with the peak baby-boom housing demand, housing prices in many areas shot up, causing an acute housing affordability problem (Dowall, 1984).

As these local growth and restrictive land-use regulations poliferated, many critics claimed that they severely impaired the operation of the land and housing markets, pushing the costs of housing beyond the reach of young moderate-income households. Figure 8.1 illustrates the deterioration of housing affordability in the US over the 1970s. It charts the ratio of the mortgage, taxes and insurance payments necessary to purchase the median-priced house in the nation (new and existing) divided by 25 per cent of median family income. At the worst point, in 1978, the cost of purchasing the median-priced single family home was 2.45 times the amount which could be 'afforded' on the median family income. The housing affordability problem was most severe in rapidly growing areas with rising land costs. The monthly mortgage payment needed to purchase the average-priced home was highest in San Diego and lowest in Detroit.

The arguments against too much restriction were compelling, at least to State and Federal Government planners, and in some states legislation was adopted to attempt to stem the application of restrictive land-use controls (Burchell, Beaton and Listokin, 1983). In the early 1980s, restrictive land-use and growth controls became a less important cause of the housing affordability problem, as interest rates rose above the 'psychological' barrier of 12 per cent, causing the demand for housing to fall. Housing starts dropped sharply, the construction industry collapsed, and the growth control debate faded. The industry came back to life in 1984, as interest rates eased, but the 'overheating' of the late 1970s, has not returned.

Land Taxation

At this point, we must mention the impact of land taxation. The most important tax on real property is the local Property Tax, and recent State legislation concerning it - notably Proposition 13 in California - has had a major impact on urban development. The Property Tax was originally the mainstay of local finance, and is still the most important local tax. In most communities, it is based on periodic assessments of the capital value of houses and other real estate, with the tax rate being determined by the local community. In the 1970s, the average property tax as a percentage of the value of single-family homes in the USA was around 2 per cent, compared with 1.5 per cent in 1962; the average tax per capita (in constant dollars) had risen by over a quarter.

Most economists conclude that the Property Tax has both virtues and weaknesses (Netzer, 1970). It is simple to collect, not easily evaded, and provides an independent source of

Fig. 8.1: Housing affordability index. 1.0 is 'affordable'

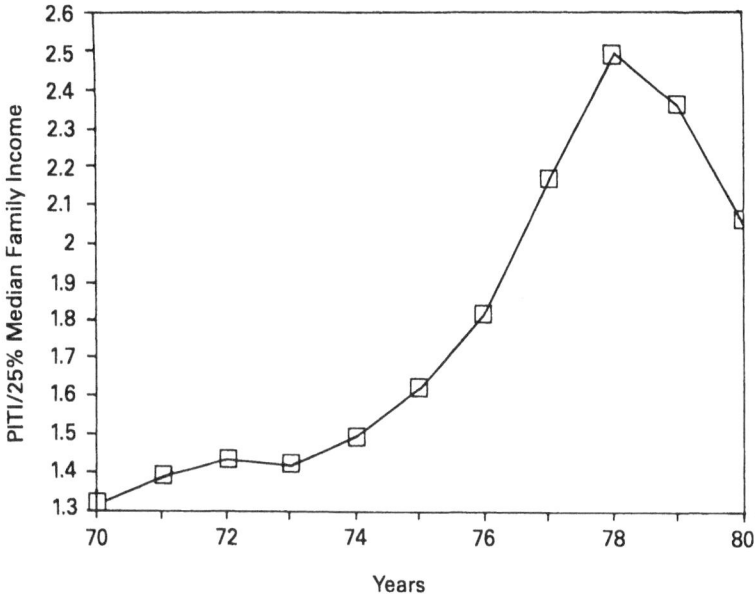

Years

revenue to local communities. On the other hand, if it rises above a fairly low level, it can give rise to inequities, since the value of a home is not always proportionate to the owner's income. It may also bear more heavily on central city districts than on the suburbs. Most economists support the development of other taxes - or 'revenue sharing' of State taxes - in order to reduce local communities' dependence on the Property Tax. In the event, little was done in the 1960s and 1970s to provide other sources of local finance.

The dependence on the Property Tax, combined with rising home prices and administrative inertia, produced a taxpayer's revolt in California. Property taxes in California, and many other states, were calculated on the basis of current market value of the property, as reflected by sale prices of comparable property. With the advent of computer-based mass appraisal, local government assessments became extremely efficient, and in California, property valuations increased rapidly, reflecting the dynamic quality of the market. Increases in assessments do not automatically trigger increased taxes, since the levied tax is a product of both the assessed value and the Property Tax rate. Local elected officials, however, did not lower the tax rates. They left the rates unchanged, and reaped enormous windfalls as local government tax revenues skyrocketed.

This situation caused widespread concern, and provided a receptive climate for a conservative populist politician,

Howard Jarvis, who launched a ballot initiative to amend the State of California's constitution, to limit the amount of Property Tax which may be levied by a community. 'Proposition 13' gained sufficient support, and was implemented in 1978.

For properties purchased after the passage of Proposition 13, the tax is limited to one per cent of the purchase price. Thereafter, as long as the property is not resold, the tax levy can increase by no more than two per cent per year, regardless of the rate of property appreciation. When the property is sold, its market value is adjusted to the purchase price, and the tax can again increase by two per cent per year. For properties purchased before 1975, the tax is based on the property's 1975 market value. This value is increased by two per cent per year. (This figure has generally been well below the inflation rate, which peaked at 13 per cent in 1980; it fell to 2 per cent in 1986 but in 1987 was rising towards 4 per cent.)

The sponsors of Proposition 13 were concerned to limit not only the Property Tax, but all taxation - and thus curb 'wasteful' public expenditure. Proposition 13 therefore prohibited increases in other State or local taxes, unless authorised by two-thirds of the legislature or electorate, respectively. Some other States subsequently adopted comparable restrictions.

The passage of Proposition 13 has led to enormous changes in local government planning. First, since taxes may not increase rapidly unless there is turnover of property, communities are carefully considering the fiscal implications of permitting new development. Residential developments, for example, are 'undesirable' because they have low turnover rates and generate large demand for services, such as schools, police and fire protection. Second, with the revenue short-fall, communities are now considering new approaches to revenue enhancement, such as selling and leasing public property to private users, and contracting-out services. The privatisation of public services has become widespread and will grow as federal subventions to local governments decline.

But the biggest impact of the Property Tax revolt has been on the financing of infrastructure. Because local governments can no longer increase tax rates without a two-thirds approval of the voters, infrastructure cannot be financed through tax increases. Thus new techniques have evolved, such as 'benefit assessment districts', exactions, impact taxes, fees and user charges. Some of these techniques are a revival of 'special assessments', which have a long history going back to the 18th century; they involved levying a charge on landowners who benefited from local publicly-financed improvements, such as flood control. The creation of 'benefit assessment districts' involves levying a charge on all landowners in a specified district; 'exactions' (known as 'planning gain' in the UK) are requirements on a developer to help finance public facilities in some way (Babcock, 1987). Some communities have imposed substantial fees on developers, or 'impact taxes' on development, such as a tax on every new bedroom built. Although all these techniques are different, they all attempt to shift the cost and financing of new infrastructure onto the developer or user. This has raised the costs of development, and it has been alleged that housing costs have been driven up as a consequence.

The infrastructure funding problem also extends to existing, especially old, infrastructure. In many cities, roads, sewer and water lines, and schools are in desperate need of repair, but (in States where Property Tax limitations have been adopted) it is extremely difficult to get voters to approve tax increases. In California, however, voters in Santa Clara and Alameda Counties have voted for increased taxes (on the sale of goods) to allow roadway improvements.

In Pittsburgh and Hawaii, the Property Tax is assessed on the value of the site rather than of the building (site value taxation). This system has enthusiastic supporters, who maintain that it has had highly beneficial effects; others, however, are unable to discern any significant benefits (Hagman and Misczynski, 1978, pp. 411ff).

Federal taxes can be summarised briefly. The USA has a tradition of treating capital gains as income, usually at reduced rates. Capital gains from real estate held for less than a year are taxed as income, the maximum rate being currently 38 per cent. If the receipts from real estate held for more than a year are re-invested in real property within two years, no capital gains tax is chargeable; thus most people selling and buying a house would not be taxed. Apart from this concession, the receipts from the sale of real estate held for more than a year will (under the 1986 Tax Reform Act) from 1988 onwards also be taxed as income. Since such gains will include 'paper profits', the new rates of Capital Gains Tax will be relatively high. There is no special tax on rises in land values, except in the State of Vermont, which has a tax on short-term gains.

New Concerns about Economic Development and a New Twist in Growth Control

Since 1982, land-use planners have shifted their attention from residential growth controls to economic development, infrastructure and financing issues. Many are now involved in partnerships linking private real-estate developers with public agencies (Levitt and Kirlin, 1985). Cities such as San Diego, Philadelphia, Baltimore, Oakland and St. Louis have provided land and low-cost tax-exempt financing to developers to build projects aimed at rejuvenating downtowns.

In some communities, office and commercial development has occurred rapidly, causing tremendous planning problems. San Francisco has adopted a growth control program for downtown commercial office buildings, while several suburban communities have imposed growth controls on commercial development (Kroll, 1985; Dowall, 1986). So have other cities including San Diego, Princeton, New Jersey and Tysons Corner in Virginia (Cervero, 1986). These new attempts at controls over commercial development are a marked departure from the residential-focused growth controls of the 1970s (Keating, 1986).

Land-use planning and policy-making are taking new twists in cities across the nation. Planning is much more connected with the private sector than it was five years ago, and new and sophisticated land policies have been implemented to respond to changing economic, social and physical conditions.

STRUCTURAL CHANGE AND THE TRANSFORMATION OF AMERICAN CITIES AND SUBURBS

The US, like Western Europe, is living through a virtual revolution, which is modifying the economic, social and spatial characteristics of our nation (Ebel, 1985; Noyelle and Stanback, 1984). The effects are felt daily: plants close, offices move to the suburbs, businesses relocate to the south and west, and technology displaces workers (Bluestone and Harrison 1982). Accelerating changes in technology, international trade, employment and resource utilisation are reshaping the patterns and processes of metropolitan development, and raising fundamental questions about the appropriateness of current local land-use planning policies and programs (Committee on National Urban Policy, 1983).

These trends are likely to continue. Further intensification of trans-Pacific trade will attract more financial and business-serving organisations to West Coast locations. Some cities will benefit from international trade - cities like Los Angeles, San Francisco and New York. On the other hand, Japan's penetration of West Coast steel markets has virtually eliminated large-scale steel production in California (Shapira, 1984). These trade shifts mean that some US firms will be hurt, and others will benefit greatly; over time there will be profound changes in the demand for commercial buildings. The loss of competitiveness in some industries and the growth in others is also likely to cause big shifts in labour markets, and indirectly affect the demand for housing. In some cases, housing demand will plummet as workers are laid off; in other cases housing demand will rise sharply, fuelled by rapidly expanding employment. Most often, these extremes occur in different sections of the nation, or in different portions of a state. However, they sometimes coexist in the same city, when the economy is shifting from manufacturing to services and finance (Mullin, Armstrong and Kavanagh, 1986). Pittsburgh is an example; its steel and manufacturing industries are sloughing off jobs at the very same time that business and financial services are expanding. Unfortunately there is little cross-over between these two industries, and Pittsburgh experiences growing prosperity amid growing poverty.

Two fundamental questions for land policy are how the newly emerging technologies will change urban development patterns, and whether we can anticipate their effects. Office automation processes - optical scanning devices and microcomputers - certainly seem likely to reshape the demand for office space. One study suggests that, as more upper-level managers adopt microcomputers, armies of clerical workers and many middle-managers will be rendered redundant (Baran, 1985), which could cause the demand for office space to plummet. However, the application of office automation technology is beginning to alter the size pattern of firms. Small firms are increasing in number, and use more space per worker; this may offset the decline in the demand for space by larger firms (Dowall and Salkin, 1986).

These technological impacts go well beyond the demand for sheer space, profoundly altering the desired location of office projects. Modern telecommunications enable office functions to be conducted in remote locations. Large financial service companies have consolidated routine clerical functions in rural or outlying suburban areas, where both labour and building costs are low. (Office space in San Francisco's financial district costs five times as much as in nearby suburbs.) This decentralisation of office activity is not new, but the pace and extent of it is (Foley, 1957). In many suburban office nodes, the pace of development has created severe traffic congestion and housing inflation problems (Kroll, 1984; Lynch, 1983; Baldassare, 1986).

Changing demands for office space will affect the demand for transportation and housing. Metropolitan areas will become less monocentric. Suburban office concentrations such as Houston's Post Oak and Greenway Plaza, Phoenix's Camelback, Washington's Silver Spring, and Philadelphia's Valley Forge now compete with their respective CBDs (Central Business Districts) (Urban Land Institute, 1985). Transportation needs will change, challenging suppliers of transportation services to find alternative modes. As smaller office centres proliferate on the edges of metropolitan areas, commuting patterns will no longer be dictated by distance to the CBD. Commuting areas will continue to push outward, as people working in the suburbs choose to live in outlying rural areas.

Although these technological changes are only beginning, there are already signs of strong office decentralisation

(Schwartz, 1984). Even in slower growth areas such as Detroit, Philadelphia and Baltimore, office development in the suburbs is booming. In 1983, 59.3 per cent of office development in all metropolitan areas took place in suburban locations, although the percentages ranged from 84 in Atlanta to 9.5 in Hartford.

Perhaps the biggest question raised by this uneven development is whether government land and housing policy, alongside private industry, is capable of providing housing and infrastructure at prices people can afford, and at locations close to where they work. This combination is proving elusive in many metropolitan areas, particularly in the rapidly growing sunbelt cities of San Jose, Dallas, Los Angeles and, more recently, Phoenix. Prices are rising fast because employment is rising rapidly, and housing demand is outpacing the construction industry's ability to increase supply - sometimes because of local land-use and environmental controls. Communities are aggressively seeking commercial and high technology development, while limiting residential development. This strategy may work for one community, but when most communities in an area follow suit, housing need and job creation become unbalanced. In other areas, growth is so fast that the most well-conceived planning process, even when coupled with an aggressive building sector, cannot meet demand.

The housing price spiral causes other serious problems. If house prices are too high, new migrants will set their sights on more distant lower-priced homes. Ultimately, development patterns become 'imbalanced', with jobs concentrated in one area and affordable housing in another, resulting in severe transportation problems.

One of the profound problems confronting anyone who attempts to design a coherent national policy for the cities of this nation is the staggering diversity of growth patterns, and the associated problems and opportunities. Policies designed to respond to problems found in one city may be totally inappropriate for different problems in other cities.

EFFECTS OF THESE TRENDS ON LAND AND HOUSING

Fast-growing housing markets face problems of land-inflation, infrastructure constraints (especially traffic congestion), planning and regulatory backlash, and affordability. Declining markets face problems of housing market softness, neighbourhood decline and suburban over-development. This range of problems reflects the divergent nature of US metropolitan development. On average, growing areas increased their population by 33.9 per cent between 1970 and 1980, and 44.2 per cent between 1960 and 1970. Between 1960 and 1980, the growing metropolitan areas had growth rates of between 276

per cent (Fort Myers, Florida) and 23.5 per cent
(Bridgeport, Connecticut). In absolute terms Houston was the
biggest gainer, with an increase of 1,662,195 people. On the
other hand, the declining and steady-state metros had a
average growth of 11.2 per cent between 1960 and 1980, and
only 0.8 per cent during the 1970s. The variation within each
group was considerable, with Buffalo losing 5 per cent of its
population between 1960 and 1980, while Memphis - although
languishing in relation to other Sunbelt centres - grew by
26.5 per cent.

The pattern of housing production reflects the range of
population growth and decline. The figures for housing stock
(Table 8.1) reveal the tremendous rate at which housing is
being added in the growing metropolitan areas located in the
South and West. Atlanta, Houston, Phoenix and San Jose
together added more than one million housing units to the
nation's housing stock in the 1970s. In percentage terms,
Fort Myers led the pack with a neck-breaking growth of 158
per cent. It was followed by Phoenix with 88 per cent, and
Boise, Idaho with 82 per cent. Housing growth in the slow-
growth metropolitan areas was much less, averaging about 14
per cent during the 1970s. Pittsfield, Massachusetts, had the
greatest percentage increase - 26 per cent - while Buffalo
had the lowest - 8.2 per cent. Growth rates thus vary
enormously between metropolises, raising divergent land
policy issues.

PROBLEMS IN RAPID GROWTH AREAS

Land Inflation

Land has become one of the most significant cost factors in
housing construction. Between 1950 and 1980, land costs as a
percentage of total housing costs increased from 10 per cent
to 25 per cent. The President's Commission on Housing clearly
illustrated the effects of substantial land inflation on the cost
of new housing construction (President's Commission on
Housing, 1982). Land costs increased by 248 per cent between 1970 and 1980 (Table 8.2). In high growth areas like
California, Florida and the Front Range of Colorado, the land
cost percentage now reaches 40 per cent. In the ultimate
land-constrained markets of Tokyo, Hong Kong and
Manhattan, the figures are 60-70 per cent.

Prices for land, (either raw acreage or finished residential lots) result from the interaction of demand factors
such as employment growth, income, immigration and the
availability of mortgage credit, and supply factors such as
the amount of raw land zoned for residential development, the
supply of residential lots, and the availability of infrastructure. In a study of land and housing markets in the San
Francisco Bay Area, Dowall (1984), examined the 'land-price

Table 8.1: Metropolitan Housing Stock, 1960-80

| Metropolitan Areas | Housing Stock | | | Percentage Change | | |
	1960	1970	1980	'60–'70	'70–'80	'60–'80
		000			per cent	
Growing Metros						
Bridgeport, CT	208,977	252,334	293,194	20.7	16.2	40.3
Poughkeepsie, NY	54,647	67,962	85,336	24.4	25.6	56.2
Bismarck, ND	16,402	19,300	30,056	17.7	55.7	83.3
Springfield, MO	48,033	59,619	83,441	24.1	39.9	73.7
Houston, TX	467,780	671,882	1,160,200	43.6	72.7	148.0
Atlanta, GA	353,567	514,761	768,209	45.6	49.2	117.3
Raleigh-Durham, NC	93,325	131,165	200,259	40.5	52.7	114.6
Austin, TX	85,079	121,922	215,739	43.3	76.9	153.6
Fort Myers, FL	21,032	42,074	108,598	100.0	158.1	416.4
Phoenix, AZ	211,865	316,989	596,049	49.6	88.0	181.3
San Jose, CA	199,922	336,192	473,500	68.2	40.8	136.8
Boise, ID	30,782	37,145	67,785	20.7	82.5	120.2

Declining Metros

Pittsfield, MA	23,464	25,758	34,710	9.8	25.8	47.9
Pittsburg, PA	740,838	788,433	873,071	6.4	9.7	17.6
Buffalo, NY	409,765	433,183	471,805	5.7	8.2	15.1
Nassau Country, NY	590,754	720,905	836,592	22.0	13.8	41.6
Sioux City, IA	39,169	39,264	45,117	0.2	12.9	15.2
Detroit, MI	1,219,938	1,393,896	1,588,119	14.3	12.2	30.2
Wichita, KS	127,278	134,371	162,676	5.6	17.4	27.8
Baltimore, MD	542,029	652,800	796,321	20.5	18.0	46.9
Memphis, TN	205,557	255,401	331,381	24.3	22.9	61.2
Birmingham, AL	225,296	253,816	324,891	12.7	21.9	44.2
Pueblo, CO	34,680	37,359	42,495	7.7	12.1	22.5
Great Falls, MT	24,086	27,131	31,885	12.6	14.9	32.4

Source: US Bureau of the Census. 1982. State and Metropolitan Area Data Book.

Table 8.2: Approximate Cost Breakdown for New Single-
Family Homes

	1970		1980		Per cent change 1970–80
	Cost $	Per cent	Cost $	Per cent	
Land	4,450	19	15,500	24	248.3
On-site labour	4,500	19	10,350	16	130.0
Materials	8,650	37	22,000	34	154.3
Financing	1,600	7	7,700	12	381.3
Overhead, profit, other	4,200	18	9,050	14	115.5
Total	23,400	100	64,600	100	176.1

Source: President's Commission on Housing

residuals' for single-family new houses. The land-price residual technique estimates land values by deducting the cost of
housing construction from the sales price of the housing,
thus obtaining an estimate of land cost plus development
profit. Over the period 1976 to 1979, increases in residual
values ranged from 64 per cent for Concord to 1928 per cent
(admittedly from an initial figure near zero) for Santa Rosa.
Those communities with restrictive land-use regulations and
strong demand had the greatest increase in land residuals.

Land constraints and rising land costs are not unique to
California. A survey has been made of land prices for
finished lots (10,000 sq. ft.) and raw acreage in thirty
metropolitan areas (Hoben and Black, 1982). Real estate ap-
praisers made valuations (for 1975, 1980 and 1985) for
middle-income residential development located within 20
minutes of major employment centres. The range of lot prices
was considerable; in 1985, from $8,750 for Chattanooga,
Tennessee, to $70,000 for San Jose (Silicon Valley). Raw land
prices showed even more variation; from $3,000 per acre in
Chattanooga to $240,000 in San Jose, with an average of
$27,397.

In 1975-80 the prices of both types of land rose faster
than inflation. The consumer price index rose by 51 per cent,
whereas average lot prices rose by 81 per cent, and average
raw land prices by 157 per cent. The largest increases were
in Western metropolitan areas such as Salt Lake City, Seattle,
Boulder, Portland, San Diego and San Jose. From 1980 to
1985, the rate of increase slowed considerably, averaging 38
per cent for lots and 27 per cent for raw land, as against
retail inflation of 31 per cent. There were also cities in which

dollar prices remained constant, or even fell. Raw land prices remained constant in Miami, and fell in Portland and Salt Lake City. Although prices in real terms have, on average, been fairly stable in the 1980s, and have fallen in many cities, there are still pockets of rapid land price inflation, where the subject generates grave concern.

Infrastructure
During the past few years, the attention of policy makers has been riveted on the so-called infrastructure problem. The most widely cited study, America in Ruins, vividly portrays a serious and deepening crisis in the dilapidating condition of our nation's roads, bridges, sewer and water treatment, and distribution systems (Choate, 1981). The study concludes that it will cost upwards of one trillion dollars to correct these defects, and that such funding should be a national priority. Another study, by the Joint Economic Committee of Congress, Hard Choices, although less passionate, arrives at similar conclusions (1984). The Committee's price tag is more modest - $1,157 billion for the year 2000 - but it estimates that the available public funds, $714 billion, will fall far short of what is needed.

While much of the infrastructure problem is the result of dilapidation, the inadequate supply of new sewerage and transportation facilities poses a critical problem in growing regions. As the California Council for International Trade points out:

'These are generally considered the 'critical' facilities without which development is unlikely to occur and, conversely, which can be used to influence the nature, timing and location of development...Lack of these facilities may prevent an urban region or local jurisdiction from accommodating growth deemed by policy-makers to be publicly acceptable or desirable.' (CCIT, 1983, p. 2).

The cost of providing infrastructure has been increasing rapidly. The case of Irvine, California, is illustrative. Recently, its Facilities Funding Taskforce inventoried the major facilities needed for additional residential development - streets, parks, schools, flood control, civic and performing arts buildings, and libraries. The costs required to serve an additional 1,000 residents, or about 35 acres, were estimated to range from $5.9 to $8.5 million, or $16,500 to $23,800 per dwelling.

While it is arguable that the level of services provided to Irvine's residents is very high, reflecting the relative affluence of the community, infrastructure costs in lower-income areas can also be high. When a community has excess capacity

171

in infrastructure, the marginal cost is low, but ultimately the excess capacity gets up, and marginal cost rises. The San Jose area is almost out of sewer capacity; parts of San Bernardino need new schools; and some parts of Orange County have streets jammed with autos. New developments south of Sacramento need schools, fire stations and other facilities, but funds are not available.

Some States, like Kentucky, recognise that infra-structure availability is an important element of economic development planning, and spend considerable funds. Other States, like Alabama, are financially hard-pressed. California, still reeling from the impact of Proposition 13, estimates that it will need to spend $80.5 billion on streets, transit, air-ports, and water between now and the end of the century. The anticipated revenues available are $48.2 billion, leaving a shortfall of $32.3 billion. Expenditure is needed for both repairs and new development; in the high-growth areas of California, the pressures come mainly from new development, as they do in similar areas of Texas, Arizona and Florida.

Inadequate expenditure on infrastructure has already caused a decline in the quality of life in high-growth areas. In Houston, where the population increased by 50 per cent during the 1970s, public service quality fell dramatically. A survey of Houstonians in 1979 revealed widespread dissatis-faction with public transportation, street maintenance, drain-age and flood control, and parks and recreation facilities (The Rice Center, 1981). These citizen concerns have not gone unnoticed by the Mayor or the City Council. In 1984 the City broke ground on a new 10,500 acre park, and major plans for highway expansion are moving forward. The re-sponse by the city to problems ignored during Houston's growth phase underscores the inevitability of providing more infrastructure. The recent shift in planning and development policy, requiring building setbacks and sign (billboard) controls, also suggests that Houston's sole reliance on markets to achieve desirable patterns of development has not worked well.

Planning Backlash

In many fast-growth areas, citizen concern about the declin-ing quality of life has (as we have indicated) led planners to implement highly restrictive land-use regulations. These controls have normally been implemented on a community-by-community basis and, where the number of 'no growth' com-munities is limited, little significant effect on housing markets has been registered. In some areas, however, enough towns and counties have enacted growth controls to create a 'critical mass' of regulation, which distorts land and housing markets – a phenomenon which has been called the 'suburban squeeze' (Dowall, 1984). In the San Francisco Bay Area, the combined play of local land-use regulations, limited infrastructure, and environmental politics has increased the cost of new single-family housing by 20-30 per cent. While this situation is unparalleled nationally, there are other 'high-growth, big-backlash' areas. In Fairfield County, Connecticut, towns have adopted large-lot zoning ordinances, making it virtually impossible for moderately-priced housing to be built. This has also begun to happen in 'Silicon Valley', California, causing some firms to move their expanded operations to Texas.

Affordability

The combined play of these problems of land price inflation, infrastructure constraints, and planning backlash have driven up the price of housing in many parts of the country. As Table 8.3 illustrates, the price of new housing across the nation has risen dramatically, but especially in the North-East and the West. Unless the growth areas can respond better to the housing needs of moderate-income households, affordability problems will continue.

HOUSING MARKET PROBLEMS IN DECLINING AREAS

While the problems of the fast-growing areas are enormous, problems of housing abandonment and under-utilisation are gripping many areas of the nation. The roots of decline lie in the dynamics of demography, employment, and income and capital movements into and out of urban areas. As Downs points out, neighbourhood decline can be caused by both the large-scale in-migration of poor households into a neighbourhood, and by its cessation (Downs, 1981). If there is substantial in-migration to a metropolitan area - as happened in Houston in the 1970s - a rapid pace of suburban housing construction may not cause the deterioration of central city districts, since the people who leave for the suburbs will simply be replaced by newcomers. If, on the other hand, there is little in-migration to the metropolitan area, but large numbers of housing units are built in the suburbs, the relatively affluent will move out of the central city neighbourhoods, accentuating their decline. Two prime examples of this condition are Pittsburgh and Cleveland. There is, however, another side to the coin. If suburban housing is expanding slowly, the housing costs that the poor must pay in the central city will be higher. There is thus a dilemma. Should house prices be kept high, raising housing costs, or allowed to fall to very low levels, with the risk of dereliction? Table 8.4 presents data on the population and housing market dynamics of growing and declining metropolitan areas. It reveals a pattern of suburban over-building and central

USA

Table 8.3: Median Sales Price of New Privately-built Single-Family Units

Year	US Total $	Northeast	Midwest	South $	West
1967	22,700	25,400	25,100	19,400	24,100
1970	23,400	30,300	24,400	20,300	24,000
1975	39,300	44,000	39,600	37,300	40,600
1980	64,600	69,500	63,400	59,600	72,300
1985	84,300	103,300	80,300	75,000	92,600
1986*	89,600	119,500	89,500	79,800	95,500

* As of June, 1986

Source: US Bureau of the Census, various years

city population loss in declining metropolitan areas. In growing areas, central city populations are more stable.

LAND POLICY PRESCRIPTIONS

In the past five years, innumerable housing task forces have been convened to assess the housing crisis, and propose solutions (Urban Land Institute, 1982). The strategies they propose for local land-use planning should be seriously considered as a basis for reform. What reforms are needed?

First and foremost, zoning needs a complete overhaul. Cities and counties have used residential zoning to limit the incursion of obnoxious non-residential uses, and to protect the social and economic status of neighbourhoods. But such zoning can be used to exclude the poor, by making housing too expensive. While there is little that can be done to quell racial and economic prejudices, at least outside the courts, planners can make zoning more sensitive, and more appropriate to changing lifestyles.

The zoning system controls population density by limiting the number of dwelling units per acre. Some control is necessary to ensure that the need for public services can be met, but population density is not solely determined by the number of dwelling units, and herein lies an area for reform. In the 1950s and 1960s, planning standards were predicated on three to five persons per unit, and seven to eight units an acre, giving about thirty persons per acre. Recently, the number of persons per unit has fallen from the three-to-five

174

range to the two-to-three range. This implies that population density per acre is falling, and that more units could be built without exceeding historial density standards. Planning studies of neighbourhoods could be conducted to determine actual population densities. Then, a 'second unit potential' could be established for determining the number of new units that could be permitted (by redevelopment at slightly higher densities) without creating demands for local services likely to exceed the available capacity (Gellen, 1985).

A major concern for residents about 'second units' is the impact on parking. An obvious way around this problem is to require additional off-street parking, as has been done in California and on the East Coast (Popenoe, 1977). In some areas, however, there is an excess supply of curb-side parking space (Schoup, 1983).

In growing areas, the problems of the high cost of land could be tackled by adopting higher density development levels, coupled with the provision of infrastructure in targeted areas; this should help to provide an ample supply of land. This targeting could be designed to minimise the conversion of prime agricultural land, and the degradation of environmental resources. To make such plans work, there must be clear and effective coordination between local governments. Otherwise, parochial interests will continue to prevail, and local land-use plans will not effectively balance the needs of housing and commercial development.

In declining areas, the rate of suburban development should be moderated, to help shore up inner-city neighbour-hoods. One proposed policy goal for State Governments and the Federal Government is to get households who leave the central city for the suburbs to pay the full social costs of coping with the poverty they leave behind. This principle has a bearing on the financing of infrastructure. Where infra-structure costs are high, and housing markets tight, the costs of infrastructure should be spread over the entire metropolitan area of the State. But in metropolitan areas with soft housing markets, households moving to the suburbs should bear the full costs of infrastructure (Downs, 1981). In such areas, a limitation on housing development in the suburbs can be beneficial (Scott, 1975). If all communities in the suburban ring of metropolitan areas of this type im-plemented growth controls, problems of suburban housing over-supply would disappear. Unfortunately, it would be extraordinarily difficult to generate unanimous support for such a program; land-use policy is a closely guarded local government power.

The directions for land policy in growing and declining areas are thus diametrically opposed. In growth areas, land policy needs to be expansive, if the problems of land-inflation are to be avoided. In declining areas, on the other hand,

Table 8.4: Housing Construction and Changes in Population and Households: Twenty Metropolitan Areas, 1970-75

Metropolitan area	1 Population change Number	%	2 Household change Number	%	3 New housing units	4 Ratio of 3 to 2	5 New housing in suburbs %	6 Central city population change %
Losing population								
New York	-412,627	-4.1	-58,215	-1.5	197,737	-	32.8	-5.2
Cleveland	-97,004	-4.7	26,315	4.0	51,247	1.9	88.0	-14.9
Pittsburgh	-79,138	-3.3	28,478	3.8	48,305	1.7	91.0	-11.8
Newark	-58,443	-2.8	-978	-0.2	30,731	-	91.5	-11.1
Los Angeles	-54,992	-0.8	109,511	4.5	208,874	1.9	63.1	-3.0
St Louis	-44,342	-1.8	29,954	4.1	73,086	2.4	95.5	-15.6
Buffalo	-22,363	-1.7	9,562	2.3	34,704	3.6	93.8	-12.0
Seattle	-17,859	-1.3	42,985	9.1	44,554	1.0	81.7	-8.2
Philadelphia	-17,109	-0.4	44,500	3.0	144,828	3.2	85.9	-6.9
Paterson	-8,350	-1.8	4,900	1.1	16,882	3.4	86.4	-6.0
Total or average	-812,227	-2.3	237,012	3.0	850,948	3.6	81.0	-9.5

Gaining population

Houston	286,931	14.4	130,477	21.4	121,028	0.9	23.9	8.3
Orange County	278,433	19.6	144,711	33.2	142,540	1.0	80.9	16.4
Phoenix	251,989	26.0	139,456	46.1	146,853	1.1	57.4	12.9
San Diego	226,729	16.7	115,400	27.3	145,111	1.3	58.9	11.0
Atlanta	194,611	12.2	76,700	17.9	163,950	2.1	89.3	-11.9
Denver	173,841	14.0	97,562	24.9	144,377	1.5	78.1	-5.9
Miami	171,689	13.5	81,800	19.1	155,383	1.9	84.0	9.0
Dallas	149,601	6.3	97,533	19.8	140,894	1.4	62.8	-2.6
Orlando	129,394	28.5	43,756	32.6	75,250	1.7	81.0	14.1
Washington	112,446	3.9	100,833	11.2	163,393	1.6	95.9	-6.0
Total or average	1,975,664	15.5	1,028,228	25.3	1,399,049	1.4	71.2	4.5

Source: Downs, Neighborhoods and Urban Development

USA

growth controls in the suburbs would help, not hurt, the central city.

REFERENCES AND FURTHER READING

Alonso, William (1973) 'Urban Zero Population Growth', Daedalus, Fall
American Law Institute (1976) A Model Land Development Code, Chicago: ALI
Babcock, Richard (1964) The Zoning Game, Madison: The University of Wisconsin Press
Babcock, R. (1987) 'Exactions: A Controversial New Source for Municipal Funds', Law and Contemporary Problems, Vol. 50 No. 1, Durham, North Carolina
Baldassare, Mark (1986) 'Suburban Industrialization: Density, Diversification, and Changing Public Attitudes', in J. Dimento (ed.), The Urban Caldron, Boston: OG&H
Baran, Barbara (1985) Technological Innovation and Deregulation: The Transformation of the Labor Process in the Insurance Industry, Berkeley, Berkeley Round-table on the International Economy, University of California
Bluestone, Barry and Harrison, Bennett (1982) Deindustrialization of America, New York: Basic Books
Burchell, Robert W., Beaton, W. Patrick and Listokin, David (1983) Mount Laurel II: Challenge and Delivery of Low-Cost Housing, New Brunswick, NJ: Center for Urban Policy Research
California Council for International Trade (1983) 'Role of the State of California in International Trade', Oakland: CCIT, 17 March
Cervero, Robert B. (1986) Suburban Gridlock, Piscataway: NJ Center for Urban Policy Research
Choate, Pat and Walter, Susan (1981) America In Ruins Washington: Council of State Planning Agencies
Committee on National Urban Policy (1983) Rethinking Urban Policy, Washington: National Academy of Sciences
DeGrove, John (1985) Land, Growth and Politics, Chicago: Planner's Press
deNeufville, Judith I. (ed.) (1981) The Land Use Policy Debate in the United States, New York: Plenum
Dowall, David, E. (1980) 'An Examination of Population Growth Managing Communities', Policy Studies Journal, Vol. 9 No. 3, pp. 414-427
Dowall, David E. (1984) The Suburban Squeeze, Berkeley: The University of California Press
Dowall, David E. (1986) 'Back-Offices and San Francisco's Office Development Growth Cap', Working Paper No. 448, Institute of Urban and Regional Development, University of California, Berkeley, March

178

Dowall, David E. and Salkin, Marcia (1986) 'Office Automation and the Implications for Office Development, Working Paper 447, Institute of Urban and Regional Development, University of California, Berkeley, May

Downs, Anthony (1981) Neighborhoods and Urban Development, Washington, The Brookings Institution

Ebel, Robert D. (1985) 'Urban Decline in the World's Developed Economies: An Examination of the Trends', in Causes and Consequences of Urban Change in the World's Developed Countries, Research in Urban Economics, Vol. 5. Greenwich, Conn.: JAI Press

Fellmeth, Robert, Project Director, Politics of Land: Ralph Nader's Study Group on Land Use in California, New York: Grossman

Foley, Donald (1957) The Suburbanization of Administrative Offices in the San Francisco Bay Area, Real Estate Research Program, University of California, Berkeley, No. 10

Gellen, Martin (1985) Accessory Apartments in Single-Family Housing, Piscataway, NJ, Center for Urban Policy Research

Glickman, Norman J. (1980) The Urban Impacts of Federal Policies, Baltimore: The Johns Hopkins Press

Hagman, Donald G. (1980) Public Planning and Control of Urban and Land Development: Cases and Materials, 2nd Edition, St. Paul, Minn.: West Publishing

Hagman, Donald G. and Misczynski, Dean J. (1978) Windfalls for Wipeouts, Chicago, American Society of Planning Officials

Hoben, James and Black, J. Thomas (1982) 'Residential Land Prices: Variations Between Metropolitan Areas', unpublished manuscript

Keating, W. Dennis (1986) 'Linking Downtown Development to Broader Community Goals: An Analysis of Linkage Policy in Three Cities', Journal of the American Planning Association, Vol. 52 No. 2. pp. 133-141

Kent, T.J. Jr. (1964) The Urban General Plan, San Francisco: Chandler

Kroll, Cynthia (1984) 'Employment Growth and Office Space Along the 680 Corridor: Booming Supply and Potential Demand in a Suburban Area', Working Paper Series, Center for Real Estate and Urban Economics, University of California, Berkeley, No. 84-75

Kroll, Cynthia (1985) 'Suburban Squeeze II', A paper presented at the Annual Meeting of the Association of Collegiate Schools of Planning, Atlanta, October

Levitt, Rachelle L. and Kirlin, John J. (eds.) (1985) Managing Development Through Public/Private Negotiations, Washington, The Urban Land Institute and the American Bar Association

Lynch, James (1983) Case Study of Fairfield County, Connecticut, Berkeley, Department of City and Regional Planning: University of California

Mullin, John R., Armstrong, Jeanne H. and Kavanagh, Jean S. (1986) 'From Mill Town to Mill Town: The Transition of a New England Town from a Textile to a High-Technology Economy', Journal of the American Planning Association, Vol. 52 No. 1. pp. 47-59

Netzer, Dick (1970) Economics and Urban Problems, New York: Basic Books

Noyelle, Thierry J. and Stanback, Thomas M. Jr. (1984) The Economic Transformation of American Cities, Totowa, NJ: Rowman and Allanheld

Popenoe, David (1977) The Suburban Environment: Sweden and the United States, Chicago: The University of Chicago Press

Popper, Frank J. (1981) The Politics of Land-Use Reform, Madison: The University of Wisconsin Press

President's Commission on Housing (1982) The Report of the President's Commission on Housing, Washington, GPO

Reilly, William (ed.) (1973) The Use of Land: A Citizen's Guide to Urban Growth, New York: Crowell

The Rice Center (1981) Houston Initiatives, Houston: Rice, Center

Schoup, Donald (1983) 'Curb Parking as a Commons Problem', Los Angeles: Graduate School of Architecture and Urban Planning, UCLA

Schwartz, Gail Garfield (1984) Where's Main Street, USA? Westport, Conn.: The Eno Foundation

Scott, Randall W. (ed.) (1975) The Management and Control of Growth, Vol. II, Washington: The Urban Land Institute

Shapira, Phillip (1984) 'California's Smokestack Industries: Plant Closures and Job Loss', Working Paper 437, Institute of Urban and Regional Development, University of California, Berkeley

Southwestern Area Commerce and Industry Association (SACIA) (1982) Report on Housing Alternatives for Middle Income Employees in Southwestern Connecticut, Stamford: SACIA

Teitz, Michael B. and Dowall, David E. (1980) Community Impact Analysis: The Palm Desert Shopping Center UDAG Proposal. Report prepared for Marshall Kaplan, Deputy Assistant Secretary for Urban Development, US Department of Housing and Urban Development, 10 April

The Urban Land Institute (1982) The Affordable Community: Adapting Today's Communities to Tomorrow's Needs, Washington: The Urban Land Institute

The Urban Land Institute (1985) Development Review and Outlook, Washington: ULI

US Advisory Commission on Intergovernmental Relations (1968)

Urban and Rural America: Policies for the Future, Washington: ACIR

US Bureau of the Census (1982) State and Metropolitan Area Data Book, Washington: GPO

US Congress, Joint Economic Committee (1984) Hard Choices, Washington: The Committee

US Department of Housing and Urban Development (1978) The President's National Urban Policy Report, Washington: USDHUD

US National Commission on Urban Problems (Douglas Commission) (1968) Building the American City, Washington: The Commission

Wolf, Peter (1981) Land in America: Its Value, Use and Control, New York: Pantheon

Chapter Nine

CONCLUSION

Graham Hallett

SIMILARITIES AND DIFFERENCES

In reading the national chapters, I have continually been
struck by the similarities of urban land problems and issues
in different countries, as well as by the differences in
national responses. There are, first of all, similarities in
urban dynamics. The dispersal of economic activity and
settlement arising from the growth of road transport, and the
accompanying 'urban sprawl' and inner-city decay, began in
the USA, but is now discernible - in varying degrees - in all
the other countries (except perhaps Yugoslavia, which will no
doubt catch up in a few years). There are also regional
changes, which cause some areas to face problems of rapid
population growth, while others face the problems of decline.
In many countries, one encounters the 'lifeboat syndrome' -
people in desirable suburban locations try to stop others
moving in. There has been a general growth of owner-
occupancy - although the USA has probably reached the limit,
and the UK may be reaching the limit of what is socially
tolerable. In all countries, there has been a switch from
greenfield development to urban renewal, and from 'comprehensive redevelopment' to renewal of a more 'cellular' kind,
involving more collaboration with private agencies. In all
countries plans take much longer to draw up and implement
than is originally envisaged - so that rigid plans tend to be
overtaken by events.

At the same time, there are political and cultural differences, which affect the policy response to these problems.
As a framework for classifying and comparing policies, an
economist is inclined to begin with the distinction between a
'command' or 'centrally planned' economy, and a 'market'
economy. However, these concepts are useful only as limiting
cases. Certain types of public activity, organised on 'non-
market' principles, are the essential basis for a market system
- if life is not to be 'nasty, brutish and short'. What Adam
Smith called 'those public institutions and those public works

which are in the highest degree advantageous to a great society' include - as an absolute minimum - a system of land registration and an independent and incorrupt legal system for supervising real estate transactions; a system for providing essential public services; and a minimal level of town planning.

Beyond these basic requirements, there are 'optional' areas of public action. Town planning ranges from the minimal level of US 'zoning' to the 'social engineering' of some of the large public housing projects. Public intervention in the land market ranges from the rarely exercised right of 'eminent domain' in the US, through the German or French participatory systems to the monopolistic arrangements for acquiring greenfield sites in the Netherlands.

One basic issue is 'whether the measures to be used are to supplement and assist the market or suspend it and put central direction in its place' (Hayek, 1960). Public authorities may make a deliberate and informed (and not necessarily unjustified) choice to substitute 'central planning' for a market system in some particular field of urban development - as the Dutch have done for 'greenfield' land acquisition, and as all town planning systems do, to some extent, for landuse. But policies have sometimes been adopted without an appreciation of their long-term consequences. A useful German distinction is that between measures which are systemkonform (compatible with the system) and those which are not, or - more narrowly - marktkonform (compatible with the market) and those which are not. A marginal tax rate of 30 per cent (real, not paper), on gains from land sales, as in the Development Land Tax in its final form, is compatible with a market system in land sales; a rate of 100 per cent, as in the 1947 'development charge', is not. A system of 'zone planning' is compatible with a town planning system based on development control; a system in which any developer can appeal to the central government against a refusal of 'planning permission' with a high probability of gaining his appeal (the current British situation) is not.

The Countries
In the spectrum of 'market economy' to 'command economy', the USA (and, since the 'Thatcher revolution', Britain) are nearest to the 'market' end. As David Dowall points out in his chapter on the USA, the provision of housing, including land acquisition, is primarily a private activity. The role of local communities was, until the 1960s, confined to implementing a very loose 'zoning' system. The general perception was that the country had endless supplies of land, and only in the 1960s did Americans begin to realise that more careful land management was needed. But however salutary the 'environmental' reaction may have been, it took some wrong turnings.

CONCLUSION

The supply of housing was restricted, which raised prices and hurt the poorest sections of society. The restriction sometimes took forms (such as large minimum lot sizes) which were environmentally perverse; one writer refers to 'the environmental protection hustle' (Frieden, 1979). There has since been a reaction, but one which runs the risk of throwing out the baby with the bath-water.

Dowall's account of 'growing and declining America' also brings out the extent to which Americans are geographically mobile. The US land and housing system copes with this mobility in some ways well, in some ways badly. It has been successful in producing large quantities of housing - of a standardised low-density type. It has not been very success-ful in ensuring that the settlement pattern in growing areas does not produce severe transportation problems; in providing essential infrastructure; and in coping with the physical, economic and social problems of declining areas.

On the other side of the Atlantic, France, West Germany and the Netherlands have a good deal in common. They all have a long-established tradition of town planning and (especially in Germany and the Netherlands) public inter-vention in the urban land market. There are, however, political differences. France has traditionally had a strong centralised state, although a process of decentralisation to regional bodies and communes has been proceeding since the 1960s. Jon Pearsall's chapter brings out the point that individual property rights are strongly defended, but that public bodies have been given extensive powers of pre-emption in order to facilitate the acquisition of land for development. Germany has a long tradition of regional and municipal independence, which was only briefly interrupted by National Socialism. Procedures for land-banking and Umlegung were developed by German municipalities in the 19th century, and have played a major role in the rebuilding of the destroyed German cities, the development of new suburbs, and urban renewal. Of the countries examined, West Germany has the most intricate and intertwined relationship between public and private enterprise in the land market.

In the Netherlands, land policy took a particular form, outlined by Barrie Needham, because large parts of the country were originally under water, badly drained, or threatened by flooding. It was necessary for municipal governments to acquire the land, so that it could be drained and prepared; this procedure has become the norm for new greenfield developments. (Whether it is technically necessary today is perhaps open to question.) The land is then disposed of, usually, although not always, on a leasehold basis. What is most striking is that such a comprehensive system of state control of land acquisition works so smoothly in conjunction with a pluralistic and fairly market-orientated system in housing and development, and arouses virtually no controversy. With the swing to urban renewal, the Netherlands has adopted a more selective approach to land purchase.

Yugoslavia might seem to be in a different category from the other Continental European countries. In addition to being 'socialist', it lags behind the other European countries in national income, and in the fact that the urbanisation which elswhere occurred thirty or more years ago is in full swing. Nevertheless, there are many similarities, especially with Germany and the Netherlands. Yugoslav Communism is a home-grown variety, which gives great freedom to local municipalities - even more than in the other Continental European countries. Moreover, as Georgia Butina stresses, the system allows both owner-occupancy and a decentralised system of housing associations. The experience with large housing estates and urban renewal is very much in line with experience elsewhere. It is also noteworthy that Yugoslavia combines its socialism with some breathtakingly 'free market' methods (putting town planning out to tender!) and with a readiness to use 'sweeteners' where compulsory purchase or the end of a lease are involved.

About Britain, it is difficult to make generalisations, except that it is a land of extremes and policy revolutions. Since the War, three controversial 'socialistic' land schemes have been passed by Labour Governments and repealed by succeeding Conservative Governments. The Thatcher Government, which came to power in 1979, initiated an equally controversial 'radical Conservative' programme - centralising political power, privatising public housing, curtailing the powers of local authorities, cutting higher tax rates, 'capping' local rates, and now replacing the rates (property tax) with a poll tax. The disbanding of public housing, tax breaks for the rich, and the neglect of infrastructure have parallels in Reagan's USA, and 'rate-capping' has a parallel in 'Proposition 13', but the USA does not display the centralised arbitrary power, both overt and covert, which is an aspect of the 'Thacherite revolution',

Some commentators have pointed out the irony that, whereas France is decentralising, Britain is creating a Napoleonic state. There is something in this, but the comparison is unfair to France. The 'Napoleonic state', as it has existed under the Fourth and Fifth Republics, was always distinct from the Government. Its administrators were an intellectual elite who did not feel themselves to be bound to any political party or ideology. Indeed, their high point was under the Fourth Republic, when there was often no Government to speak of. The various official bodies concerned with urban development may at times have been patronising to the 'Clochemerles', but they were never deliberately used to emasculate them. The current British centralisation thus lacks the non-partisan, philosopher-king character of the 'Napoleonic state'.

LAND PRICES

In comparing national experiences, I have also been struck by similarities in the movements of land prices, and in the political reactions to them. The popular complaint is of 'ever-rising land prices', but what has actually happened is more complex. During the housing boom of the 1950s and 1960s, land prices rose, in real as well as in nominal terms. This was, however, a catching-up after the fall of the inter-War period. In Britain and Germany, at least, land prices in city areas did not, in real terms, reach the pre-1914 level. After 1973, there was a sharp fall, followed by a recovery in which average figures concealed large variations between regions and city districts. Real prices have risen in booming cities like London, Munich or Phoenix, but not in declining cities. Prices in suburban areas (which had been low) have risen, but prices in declining inner-city areas (which had been high) have often fallen. There is thus a tendency towards an evening-out of prices.

In the not-very-long run, there seem to be self-correcting mechanisms in land prices. Rises (in real terms) in the general price level tend to be followed by falls, and districts with exceptionally high prices are usually heading for a fall. As a percipient economist wrote in 1905:

> '... the course of land value seems to "crook and turn upon itself in many a backward streaming curve"' (Edgeworth, 1925).

This phenomenon does not justify complacency or laisser faire. There is still a case (which Edgeworth supported) for some form of taxation of land values, and there is still a case for public intervention in the land market, for three purposes; to assist low-income groups who are being 'squeezed out'; to establish or preserve certain types of land-use which are 'uneconomic', but which are desirable from a wider or longer-term viewpoint; and to 'smooth' the supply of land. A historical perspective on land prices does, however, justify a more phlegmatic reaction to rises in land prices than usually prevails. The rise in prices of the late 1960s caused an outcry against 'speculation', which in some cases led to taxes which came into effect only when conditions had altered (French TLE, German Ausgleichsbetraege). There are, however, differences in attitudes and institutions. In the USA, 'unearned increment' seems to arouse no concern at all. In France, the general feeling is that landowners are entitled to make some money from their land, but not to hold the community to ransom; hence the emphasis on pre-emption. In Britain, Labour Governments have been obsessed with creaming-off 'unearned increment' to the exclusion of all other considerations. The Minister in charge of the 1965 legislation even insisted that, 'The Betterment Levy is not a tax' and on this ground refused to consider amendments which would have made it a more acceptable and workable tax (Hallett, 1977, p. 131). In Germany, the initial outcry against 'speculation' was similar to that in Britain, but the legislative process was so slow and thorough that, by the time all the difficulties of a tax on development value had been rehearsed, the boom had passed - which could be seen as a justification for 'checks and balances'.

LEARNING FROM EACH OTHER

Can the countries in our study learn from each other's experiences? This question raises complex methodological issues, which are addressed in a recent book (Masser and Williams, 1986). It is not just that legal, governmental and social conditions vary from country to country. There is the deeper problem of bringing into the open the possibly unconscious intellectual and cultural assumptions made by researchers, and of separating out those elements in a land policy which are inextricably linked with national and cultural characteristics, and those which could usefully be transferred to another nation or culture. As Masser points out, '... the relative lack of research on the transfer and diffusion of planning experience is particularly surprising' (Masser and Williams, 1986, p. 165). Let us hope that such research is supported. For the moment, I would answer the question posed at the beginning of this paragraph as follows: 'In some senses, yes; in the sense of discovering whether a particular land policy should be adopted, only with severe qualifications'.

The point can be illustrated by looking first at what the USA might learn from Europe - which also tells us a lot about Europe. It might be convenient to begin with two comparative studies by Americans published a few years ago (Strong, 1979; Lefcoe, 1979). Ann Strong and George Lefcoe are 'liberals', in the American sense, who are conscious of the problems of dereliction, gross inequality, and 'sprawl' associated with the US land and housing system. They therefore were interested in looking at more 'statist' systems, to see if they offered solutions. Strong gives an account of landbanking in Sweden, the Netherlands and France; reaches a favourable verdict on the results in these countries; and urges the USA to go and do likewise.

Lefcoe, in a survey of the Netherlands and Germany, is cautious to the point of agnosticism.

CONCLUSION

'Transnational comparisons may seem disappointing if we begin with the expectation of assistance in resolving major political questions, such as whether local governments should become major land developers in the United States. If American cities entered the land development business to provide a greater concentration of land uses at somewhat higher densities than the pattern we call 'sprawl', they would likely fail unless they possessed monopoly or oligopoly powers over the land market, and had authority to provide industrial sites as well as housing tracts ... From a study of Dutch development practices there is little basis from which to assess what else might happen - for better or for worse - as a result of municipal intervention at the urban fringe in this country' (pp. 145-7).

This caution is not because Lefcoe was not impressed by Dutch (and German) achievements. He was profoundly impressed by their success in building cities which were more attractive, and safer, than US cities, but he makes two points. Firstly, other factors besides land and housing policy are involved.

'By and large none of the above four countries (UK, Sweden, Germany and the Netherlands) has had so much unemployment, such vast racial and ethnic diversity, as limited a public investment in civic design and in facilities like parks and playgrounds, and so callous a system of public transport for those who cannot afford private automobiles' (p. 41). (One might add, 'and such a widespread availability of hand-guns'.)

These differences have been important, and some still are, although they are no longer so sharp. Unemployment in the UK and the Netherlands is now higher than in the USA and, in the field of public investment and public transport, Britain has moved markedly towards the US situation. Indeed, in one respect, Britain has out-done the USA. US cities still have subsidised municipal bus services. In Britain (under the Transport Act, 1985), local authorities have been forced to privatise their bus services and withdraw subsidies. The Government maintains that there will be an improvement in services. Initial evidence is of rises in charges and cuts in services, but it is too soon to make a final assessment of this unique experiment (Stanley, 1987). The general point remains, that public policies in fields other than land and housing policy have an important influence on the urban condition.

The second point made by Lefcoe is that a crucial role is played by the political and administrative system, within which specific land policies are operated.

'The crucial difference between Dutch public sector land development and that which takes place in California (and elsewhere in the United States) is that Dutch programs are an integral part of a planning system that sets operationally significant priorities as to how land should be developed. Californian planning systems have been mostly preoccupied with minimizing the nuisance-like impacts of one land use on another, and with trying to prevent one governmental agency (the road builders, for instance) from undermining the aspirations of other local agencies (such as the park department). While an American city plan often embodies aspirations – for improved housing, less commuting, and the like, the planning apparatus has no effective means of implementing its stated goals.' (p. 133)

This pinpoints a fundamental difference. In the USA, suburban land is developed privately, predominantly in detached houses for the better-off half of the population. The poorer half has to buy houses in older areas which are being abandoned by the better-off, and so have fallen in price. A 'hand-me-down' arrangement of this type can in some markets – e.g. used cars – benefit both parties. Even in the housing market, it can to some extent serve a similar purpose. However, an excessive concentration of poor people and racial minorities in areas facing problems of decaying housing, combined with a breakdown in law and order, conventional morality, and family relationships, gives rise to the 'ghetto'.

In the Netherlands and Germany (and to some extent France and Yugoslavia), local government agencies organise land for development. They make provision for lower-income housing, which is then provided by non-profit housing agencies, using public money. However, as Lefcoe points out, these policies have their costs. Local authorities are financially supported by the national government far more than US cities, and taxes and (sometimes) house prices are higher; would Americans be prepared to pay these costs? Moreover, the Dutch (in particular) are used to local authorities acquiring large areas of land, and drawing up development plans, which are reviewed in toto by local politicians. In California, the role of local authorities has traditionally been more fragmented.

Lefcoe concluded that, if local authorities in the USA embarked on land acquisition, under existing conditions, and were confined to a small part of the land market, they would run grave financial risks. These would arise from any attempt to cater for a wider social mix than normal, or to produce anything different from the normal housing layout. Merely buying large areas of land for development is no guarantee of large financial gains – a point confirmed by Needham's account of the recent financial difficulties of some Dutch local authorities.

My own assessment is that, as Lefcoe argues, public participation by public agencies in the US land market would have to be on a fairly large scale, including profitable commercial activities, and amply supported by Federal or State funds, including housing subsidies, if it were to be successful. These conditions are not likely to be met at the present time. The wider conclusion is that there are limits to an 'à la carte' approach. If one wishes to emulate the achievements of another country, one may have to adopt more than the technicalities of land policy.

THE US ZONING SYSTEM

Having examined the scope and limits of international emulation, let us first elaborate David Dowall's conclusion, in his chapter on the US, that, 'First and foremost, zoning needs a complete overhaul'. To appreciate the implications of such an overhaul, it is necessary to understand the origins of US zoning (Delafons, 1969) and the role of law in American planning.

We can distinguish three approaches to law and planning; the Continental European (in particular, the German), the British and the American (specifically the Californian). In Germany, town planning has developed as a branch of administrative law, under which individuals can appeal to the courts if they believe that they have been treated improperly by officials. Thus a property-owner adversely affected by a town plan can appeal on the grounds that proper procedures were not carried out, or that his treatment was discriminatory compared with comparable property-owners. He cannot, in general, appeal against the substance of a properly-prepared town plan. Britain took a different route in 1947, when it introduced a town planning system which was largely outside the law, in the sense that the local planning authorities had complete discretionary power to give or withhold 'planning permission', subject to an appeal to the Secretary of State (who also had complete discretionary power).

The US zoning system was originally derived from the German, with its emphasis on 'the rule of law' and appeal to the courts (Logan, 1976). In the adaptation, however, it tended to disallow the mixed uses which had been - and still are - permitted in Germany; the 'deadening' consequences were later deplored (Jacobs, 1962). Since the Second World War, moreover, the 'rule of law' has acquired a new meaning. The Americans have become the most litiginous people in the world; vast numbers of lawyers are eager to assist anyone (on a payments-by-results basis) to divorce his wife, sue his doctor, or take the local authority to court. In the field of planning, matters which in Germany would normally be decided by local authorities (on advice from the planners, and after extensive public consultation) have (especially in California) been the subject of private lawsuits. The California Environmental Quality Act, of 1970, for example, provided that Environmental Impact Reports had to be prepared for large industrial developments – a very reasonable provision. But the courts ruled that EIRs also had to be provided for housing developments; since the EIRs could be challenged in the courts, the outcome was a large number of lawsuits, nearly always brought by opponents of development. In this way, the admirable objectives of the 'environmental' movement led to the very questionable 'suburban squeeze'.

California's planning laws also stipulated that zoning and subdivision approvals had to be consistent with a 'general plan', which must be 'internally consistent'. The general plan is comparable to British 'Structure Plans', German 'land use plans' etc. All these plans have had their problems, but there is a peculiar feature of the Californian system. Any citizen can file suit against any community for non-compliance with general plan requirements. The threat has made planners extremely cautious in their efforts to meet legal requirements.

'Largely as a consequence, general plans and EIRs are now longer and harder to understand, thickness is now substituted for meaningful content, critical priorities are obscured in the quest for legal adequacy, and creativity is stifled by the checklist mentality' (Topping, p. 181).

Various ways have been suggested for limiting the amount of potential litigation. One is that appeals should be concentrated in a special court, on the lines of the Land Use Board of Appeals initiated in Oregon in 1979. More radical proposals are that the scope of appeals should be limited, so that the courts are more concerned with whether local communities made reasonable decisions, in good faith, and according to prescribed procedures, rather than with substantive issues. Another proposal is that only persons directly affected by a proposed development should have a right of appeal. Most experts also favour a simplification of the legal requirements, greater emphasis on priorities, more continuous monitoring and revision of general plans, and more local experimentation.

The Rule of Law

The basic difficulty seems – to an outsider who supports 'the rule of law' and would like to see more of it in Britain – to be that, in the USA, the concept has been extended beyond its legitimate sphere. It is one thing to provide appeal to the courts to ensure that individuals are treated in an even-handed way. It is quite another to allow any citizen to challenge in courts the specific way in which local authorities have interpreted general planning objectives. Problems of this type are not confined to urban development. American doctors (and ultimately patients) have to pay astronomical premiums for insurance against 'malpractice' suits, and major consider- ations in diagnosis and treatment are now the implications for possible litigation.

The US situation provides an example of the methodo- logical problems involved in the cross-national transfer of land policies. When it became known a few years ago that the European Commission was proposing a procedure for the environmental assessment of development projects, some experts objected on the grounds that this would generate excessive litigation, as it had done in the USA. In fact, the litigation was more attributable to the USA's litiginous culture than to any intrinsic feature of environmental assessment reports.

The Problems of Decline
We have concentrated on the Californian problems of rapid growth; these problems are not, however, universal in the USA. In declining areas - as Dowall points out - a different approach is needed. Urban renewal needs to be accompanied by restriction on 'greenfield' development. Similar arguments have occurred in British cities, where critics have argued that it is misguided to allow large-scale suburban development when the 'inner city' is decaying. The US, however, has a special problem - the large number of separate local govern- ments in most conurbations. This fragmentation makes it difficult to stop development when no powerful local interests are against it. Dowall reaches the rather pessimistic con- clusion that coordination is needed, but would be extra- ordinarily difficult to achieve.

The argument for coordination echoes the case - often made in the USA in the 1960s - for 'metropolitan' government of conurbations. But this idea did not generally take root. Although the absence of metropolitan government is regretted by some, others argue that fragmentation has its virtues, and question whether large-scale planning is workable - at least in the USA. Lefcoe (with specific reference to California) has proposed abandoning the 'general plan' in favour of more localised improvements, and the preservation of open space through its acquisition and management by State or local authorities (Lefcoe, 1981). On this view, the multiplicity of local governments (no less than 85 in Los Angeles County) should be regarded as a virtue rather than a defect, because of the competition it offers. A variety of approaches, some of which will succeed and some of which will fail (and be abandoned), is more likely to work than the attempt to impose a large-scale 'plan'. In the USA, a case can thus be made for metropolitan government and for diversity. Britain has perhaps given itself the worst of both worlds, by abolishing metropolitan governments, without allowing local governments the autonomy enjoyed by US communities.

The problem of restricting suburban growth in declining American metropolises, however, goes deeper than the structure of local government. Are Americans prepared to accept restrictions on their right to develop their land, especially if there is no compensation if permission is refused, and (often) no tax when permission is given? This is the issue (known, somewhat incorrectly, as 'betterment') which the British Labour Governments tried to tackle - not very wisely or successfully. A more knowledgeable and rational set of proposals for taxes on 'windfalls' and compensation for 'wipeouts' was put forward by the late Don Hagman (Hagman and Misczynski, 1978). A more piecemeal approach might, however, be more feasible. In specified areas, landowners could be offered compensation if they refrained from development, and also agreed to specific 'environmental' obligations - allowing some public access, maintaining woodland etc.; this already happens in some European countries.

A 'reform of the US zoning system' is thus far more than an administrative matter; it raises sensitive issues of a wide-ranging kind. The prospects for a widespread reform are not at the moment bright, and the impetus for any change is unlikely to come from the Federal Government. There could, however, be - and there have been - developments at the State or local level.

The USA's Unknown Neighbour
Although US academics have made several studies of European land policy, they have ignored the most obvious comparison. Canada (outside Quebec) is 'American'; indeed, in some city districts a visitor would be hard put to tell which country he was in. But Canadian cities do not have the huge, decaying, racially-segregated, dangerous, central-city districts of their US counterparts. Among the causes of this difference, one would have to include the greater racial diversity of the USA. There are, however, significant differences in land policy and local government. The large Canadian metropolises - unlike their counterparts in the USA - have metropolitan govern-ments, on top of community governments, which are respon-sible for general land-use and transportation policies (Norton, 1983). Several Canadian metropolitan authorities and provinces also maintain large land-banks (Roberts, 1977). Do these differences contribute to the greater 'livableness' (although some Americans might say dullness) of Canadian cities? The Canadian discussion on land policy has ignored the US com-parison almost as much as vice versa. One of the few

available comparisons makes the point that higher Canadian land and house prices (in the early 1970s) are a 'good thing', since they reflect higher environmental quality (Goldberg, 1977). Goldberg puts forward the, at first sight paradoxical, argument that the Canadian superiority is the result of <u>less</u> government intervention.

> 'Canadian cities are compact, continually being renewed by private homeowners and investors, and are stable, slowly evolving entities. US cities, on the other hand, have been bombarded during the past 30 years by a series of 'facilitating government policies' such as urban renewal, freeway construction and public housing ... Such 'facilitation' of urban development in the suburbs and urban redevelopment in central areas has had an incredibly destabilising impact on private investment in US cities, on the composition of central city residents, and the quality of the urban environment ... The British North America Act (the core of the Canadian Constitution) effectively prevents the Government of Canada from taking any direct role in urban development. As a result, Canadian cities are not dotted with Federally built highways, public housing projects, and urban renewal.'

This may not be the whole story, but it is certainly true that Canadian local government is even more independent of 'central' government than US local government. (We had hoped to include a Canadian chapter in this book; there is certainly a useful Canadian-US comparison to be written.)

WHAT CAN BRITAIN LEARN FROM OTHER COUNTRIES?

When we cross the Atlantic, it is probably best to confine bilateral comparisons to the lessons Britain might learn from other countries. In terms of specific land policies, these countries are more likely to be European countries than the USA; as Dowall's chapter makes clear, the USA is not the Utopia of New Classical theology. But much can be learned from the USA's wider political system - the independence of States and communities, which allows local experimentation, and the 'grassroots democracy' which throws up groups of people, who work together - and put up large sums of money - to improve 'their town'.

As regards policy instruments, there are two fields in which new ideas are certainly needed in Britain; land acquisition and development control. Although the land acquisition schemes of 1965 and 1975 were repealed, the Urban Development Corporations are, in a more limited way, entering the same field, and proposals have been made for 'land agencies'.

It therefore seems worth looking at the experience of other countries - and land acquisition systems are reviewed below. Barrie Needham makes an eloquent defence of the Dutch system, but there must be doubts whether the 'traditional' Dutch system (used for greenfield development) would be an appropriate goal for Britain in the 1990s. Even in the Netherlands, a less comprehensive approach has been adopted in urban renewal. For Britain, a system which made a limited use of public land acquisition - but operated with the continuity shown in the Netherlands - would be a feasible goal. The German and the French systems - comprising land-banking, pre-emptive purchase, and the kind of 'good offices' exemplified by Umlegung - contain much that could be considered for adoption. But both of these national systems have been built up over the years, on the basis of strong, independent local authorities (especially in Germany) and a partner-like relationship between private developers and local authorities. It could not be transplanted to a situation in which there is deep-rooted suspicion and antagonism between the central government and local authorities, and between (some) local authorities and developers. The conclusion would seem to be that Britain's prime requirement is an effective, independently financed, and 'legitimate' system of local government. Without such a basis, emulating French or German land acquisition methods is unlikely to achieve much.

The issue of development control is even more pressing, since the system created in 1947 is being steadily undermined. The unlimited discretionary power granted to the local authority, and on appeal to the central government, has been used in ways which were never envisaged by the framers of the 1947 legislation, which was based on the assumption that local plans would be fairly clear-cut, and that appeals would be the exception. The system in fact worked reasonably well as long as 'gentlemen's agreements' were observed, i.e. as long as local authorities made detailed decisions in an even-handed way, in line with published plans and guidelines, and the Secretary of State acted in a quasi-judicial manner, allowing appeals only when the refusal by the local authority was clearly not in line with local plans, or with 'natural justice'. These 'gentlemen's agreements' - like others - are now being increasingly abandoned. There have been complaints by developers that some Labour authorities are hostile to any private development. Equally significant, however, has been the change in the treatment of appeals under the Thatcher Government. Successive Secretaries of State have allowed developments which were clearly not within the terms of local plans. Large developers who are refused permission now appeal as a matter of course, and local planning (on large developments) has been reduced to a charade.

Since it unlikely that the whole country can be satisfactorily planned from Whitehall, there would now appear to be two courses open. One is to virtually abandon development control - as advocated by some Conservative intellectuals. This would be to adopt the policy on which the city of Houston once prided itself, but about which it is now having second thoughts (Siegen, 1972). The other is to move to a more 'zonal' and 'legalistic' system, which would allow developers a fair amount of freedom - possibly more freedom than in the current German or Dutch systems - within defined parameters. A study of the Dutch, German and French planning systems indicates that, in practice, some measure of planning discretion is still needed, but that a less discretionary system than the current British one is feasible, and can indeed work rather well. These systems, however, are based on a reasonable measure of consensus, and effective, independent local government. Political attitudes and institutions are, once again, more important than the technicalities of land policy.

ASSESSING NATIONAL POLICIES

Are there any general conclusions - in spite of all the qualifications about international transfer made above - which can be drawn from the experience of the countries studied? In attempting to answer this question, I shall, following the discussion in Chapter 2, apply the following criteria to specific policies;

(a) Do they facilitate an adequate supply of building land for housing in aggregate, and access to an acceptable level of housing for the poor?
(b) Do they facilitate good town planning and 'livability'?
(c) Do they provide for a reasonable level of taxation of the gains from the ownership of real property?

These criteria can be applied to;

1. Public land acquisition
2. Land taxation
3. Development control and planning.

Public Land Acquisition
The most extreme form of public land acquisition is the Dutch system, under which the local authority acquires all land for 'greenfield' development. It certainly solves the 'problem' of 'unearned increment' on virgin land, and works well in terms of planning, although it would be hard to argue that the Dutch have done notably better than the Germans, whose intervention in the land market is more partial. It is

significant, however, that other countries have not emulated the Dutch. Moreover, the emphasis in the future is likely to be on urban renewal and infilling loosely developed areas rather than on large-scale 'greenfield' development. In the approach to these problems, the Dutch and German approaches seem to be converging.

One conclusion which can certainly be drawn from the Dutch and the German chapters is that it would be most unwise today for public authorities (in the countries studied) to engage in large-scale public land acquisition in the hope of financial gain. Some of the long-term Dutch and German land banks have done well in financial terms, but the short-term risks are high. At present, the demographic trends cast doubt on whether rises in land prices in real terms will be the norm over the next generation. Public land acquisition should therefore be directed towards town planning objectives, the preservation of open space, and the provision of 'social' housing, rather than financial gain.

Given the trend to urban renewal, German experience merits careful study. Germany has, over the past sixteen years, carried out the most extensive and - many observers think - the most successful urban renewal programme in the countries studied. The participation by German local author-ities in the land market has proved to be very effective in this type of development. It has produced well-planned towns without the drabness and uniformity found in some of the larger public housing developments in the Netherlands, France or Britain.

German policies include Umlegung and land-banking by local government. Germany has also made extensive use of quasi-public agencies (e.g. the Sanierungstraeger in urban renewal). Such an arrangement can be convenient for small local authorities. It can also sometimes be expedient for surveys and consultation to be carried out by an organisation which is perceived to be independent. However, these organ-isations are responsible to the local authority, and operate under its supervision. There is no parallel with the British 'Urban Development Corporations' (although there is with a few cases in which organisations have been set up under the aegis of local authorities, e.g. the successful Byker re-development in Newcastle-upon-Tyne).

Another factor in the German system is a method of housing finance which has encouraged the emergence of relatively small units of ownership and management in rented housing. The outcome has been a 'fine-grained' urban pattern, without the large differences between city districts, or between owner-occupied housing and public housing, which are so striking in Britain and the USA - and were also a feature of the ZUPs in France. In recent years, however, France has avoided the failings of its post-War programmes; its system of pre-emptive rights - together with a range of quasi-public organisations not dissimilar to the German - also contains many features which could be adopted in other countries.

It seems reasonable to conclude that some intervention in the land market by a local public agency is desirable in large developments. With large greenfield sites, it is feasible - as the Dutch have clearly shown - for a public agency to acquire the whole of an area to be developed, provided that it merely acts as an intermediary for subsequent owners. But for a public organisation to acquire a large area, develop it, and subsequently manage the housing, appears to be a virtual guarantee of monotony, social segregation and poor maintenance. The best guarantee of 'livability' is a development process which results in a mixture of tenures and, within reason, of uses.

The Leasehold System

In addition to the acquisition of land with a view to development, there is the question whether the land should be sold freehold, or whether the state should remain the ground landlord, under a leasehold system of some type. The 19th century 'land reformers' attached considerable importance to local authorities (or non-profit organisations) remaining the ground landlord, so as to recoup some of the long-term rise in land values, and to retain some legal control over development. The system has been widely adopted in the Netherlands and Yugoslavia, and appears to work well. In Britain, the leasehold system tended to be discredited by an inequitable provision that the entire building reverted to the ground landlord at the end of a lease, and by the fact that the ground landlords were often private individuals who exploited their position. In Germany, it went out of favour as a result of the hyper-inflation of the 1920s. In France and the USA, it never became popular.

The leasehold system has considerable potential for combining 'the public interest' in land with private initiative. National experience, however, suggests that there is considerable institutional inertia. If leasehold (with reasonable provisions) becomes the norm, it is accepted; if it is unfamiliar, it makes little headway.

TAXATION: GENERAL PRINCIPLES

In the 1980s, it has become acceptable, and even fasionable, to advocate tax cuts for the rich. The 'supply side' case for cutting taxes takes two forms;

(a) the alleged immorality of progressive taxation and the undesirability of public expenditure, as

stressed by Friedman and Hayek,
(b)   considerations of 'incentive'.

The first is simply a value judgement, on which anyone's view is as good as anyone else's. I myself find the argument unconvincing; I can see no objection to moderately progress- ive taxation, up to a maximum rate of around 50 per cent, and would judge public expenditure on its merits. The second argument applies only to taxes which involve high marginal tax rates; it applies to some tax levels, but not to others. The moral case for redistributive policies - both in the tax/benefit system and in housing and land policy - remains, but needs to be rethought in the light of current conditions.

Inequalities in wealth and income have increased in the 1980s, in the countries studied, although to differing de- grees. This inequality is both reflected in the housing market and increased by some characteristics of that market - in particular the growth of a 'mono-tenurial' (owner-occupation) system (Kemeny, 1981). Inequality in 'developed' countries, however, takes a different form from that which it took in the past - and still does in many poor countries. In the past, there were a few very rich people in a sea of poverty. To- day, the inequality is at the bottom rather than at the top (Brittan, 1973). Most people in the countries we have studied are near the average (in terms of income and wealth, and housing provision) but a small minority is well below it. It is this minority - deprived of both income and cheap housing - which has suffered most from the depression and New Right policies. This new 'proletariat' lacks political power, both because of its small numbers and because most of its members lack the skills necessary to exert political influence. Since the condition of the poor and disadvantaged does not sway elec- tions, a change of public policy can arise only from a belief that their condition poses a threat to society in general (the riots in the USA in the 1960s and in Britain in the 1980s had an effect) or from compassion. The latter played a significant role in the massive Victorian programme of social reform and 'infrastructure provision', but is not so apparent in current exponents of 'Victorian values'.

If there is the political will to use public policy to benefit the poor and disadvantaged, the question arises whether the policy should take the form of income support, or the improvement of skills and living conditions. Both policies, however, cost money. There is a recurrent tendency to believe that a redistribution of public expenditure on these lines can be achieved merely by 'squeezing the rich', but this belief is not borne out by experience. Very high marginal tax rates have been shown not to produce much income. A more effective alleviation of poverty would have to involve a (small) increase - compared with what it would otherwise have been -

in the total amount of tax paid by the moderately affluent, who make up most of the population.

Such a move would be against current trends. The Governments of the USA and Britain have led the way in giving preference to tax cuts over the alleviation of poverty and the maintenance of public services. But the issue may well be revived in the 1990s, when Reagan-Thatcherism has run its course. If the pendulum does swing in this way, the question of the taxation of wealth, and in particular of the increasing amount of personal wealth embodied in housing, will have to be re-examined.

LAND TAXATION

In the taxes which bear on real estate, we can distinguish between general taxes, some of which may apply to real estate, and specific 'land taxes'. When using the touchstone of equity, the incidence of the whole array of taxes and allowances should be considered. The land taxes can be further judged in relation to, (a) their effect on the supply of building land and (b) their suitability as a means of financing local government.

The general taxes include Income Tax, Capital Gains Tax and Wealth Tax. Income tax is levied in some countries on certain types of real property sales (e.g. in Germany on land sales by farmers). When marginal rates of Income Tax are over 80 per cent – as they were for high earners in Britain under the 1974 Labour Government – they are almost certainly harmful and self-defeating. There is, however, very little evidence that tax rates in the 25-35 per cent range have this effect. Britain and the USA also have a Capital Gains Tax. When this tax was introduced by the 1974 Labour Government in Britain it was at effective rates of over 80 per cent (as distinct from the nominal rate of 30 per cent) because it was non-indexed at a time of high inflation; most of the 'gains' were merely on paper. The outcome was massive tax evasion. The tax has since been partially indexed, at the cost of considerable complexity; it applies to second homes but not to a 'principal residence'. The USA still has a non-indexed Capital Gains Tax, at a rate which has recently been increased, and may give rise to problems.

Because of the defects of a Capital Gains Tax, some public finance economists recommend a low, widely spread Wealth Tax, on German lines, as an alternative (Sandford, 1975). A more radical reform would be to replace both Income Tax and a Capital Gains/Wealth Tax with an Expenditure Tax. This would mean that all receipts, from both 'income' and 'capital', would be taxable, but with an offset for investment (Meade, 1978). In other words, someone selling a house and buying another house, or shares, of equal value, would pay no tax, but someone selling a house and not reinvesting the proceeds would pay tax on the proceeds. The Meade Report makes a powerful intellectual case, but seems to have fallen on stony ground. In the absence of such a radical reform, it is even more necessary to scrutinise the impact of tax allowances for housing, and property taxes (or their abolition), on income distribution.

The Property Tax

The specific 'land taxes' can be divided into;

    (a)   the property tax.
    (b)   'betterment taxes'.
    (c)   infrastructure charges etc.
    (d)   taxes on development value.

The oldest, and still the most important land tax (in terms of revenue), is the property tax ('rates', Grundsteuer etc.), consisting of an annual tax based on the value of the property. It was originally the only source of local taxation, and still has virtues. It is based on what is becoming the most important component of personal wealth – real property – and it does not have the disincentive effects of taxation (at high marginal rates) of income, capital gains, or development value. There has, however, been strong political pressure to hold down the property tax, either by resisting revaluations (Germany, Netherlands, France); by imposing a limit on the tax level (Proposition 13); or even by abolishing it altogether (Britain). Any local government system which relies heavily on the property tax is in danger of becoming inadequately funded, and of either losing its independence (Britain) or being unable to provide essential public services (California). There is nevertheless a strong case for retaining the property tax as an element in local finance. It is cheap to administer and certain in its application; it can, within limits, be altered locally without causing problems; and – in a rough-and-ready way – it functions as a wealth tax. Considered as a 'tax on housing' the property tax is quite high, as indirect taxes go. If the 'imputed rent' of housing is taken as 3. per cent of capital value (the average real return on government securities), a property tax rate equal to 1 per cent of capital value can be considered as a 'sales tax' of 33 per cent. Experience suggests that the property tax arouses opposition if it goes much above this level.

Site Value Taxation

An alternative form of the property tax is site value taxation (Prest, 1981). Under this system, the tax is based, not on the value of the property, but on the value of the site. Since the same total amount of money would have to be raised for the local authority, the difference in assessments would, in most cases, not be all that great. The main argument for site value taxation is that owners would be encouraged to develop their sites to the full. Vacant land in urban areas should certainly bear some kind of tax; the British and French systems are defective in this respect. In some cases, however, a site value tax could be undesirable or inequitable. It would not necessarily be desirable to tax golf courses on their value as building land - if one wished to have any golf courses. Nor would it necessarily be equitable to tax someone living in a small house on the basis of what the land would fetch if sold for offices. A priori argument would thus suggest that a change to a site value basis would have advantages and disadvantages, and is not a crucial issue; the experience of Pittsburgh and Hawaii in the USA, and of the Yugoslav 'split' system seems to confirm this Laodicean attitude. Of course, if a change to a site value basis would prevent follies like 'Proposition 13' or 'the community charge', it should be welcomed, but this seems doubtful. The 'rate-payers' revolts' appear to have arisen as a result of an over-reliance on the property tax as such, rather than of its method of assessment.

A Tax on Imputed Rent

Another tax, which used to be imposed by several countries (Britain until 1962, Germany until 1987), and is still imposed in the Netherlands, is Income Tax on the 'imputed rent' of owner-occupied housing. On grounds of tax neutrality bet-ween owner-occupied and rented housing, and between investment in financial assets and in housing, the case for such a tax is strong. But as the number of owner-occupiers has grown, so has the political pressure to abolish it. In Britain, the abolition of 'Schedule A' in 1962 opened a Pandora's Box of fiscal problems.

In Germany, the tax treatment of owner-occupancy up till 1987 was somewhat similar to that in Britain before 1962 (although the German treatment of the letting-off of part of a house, or of dwellings in a multi-dwelling building, was more favourable). The new German system substitutes a fixed 'once in a lifetime' subsidy for interest relief. This avoids the distortion of the capital market which has become so apparent in Britain. In broad terms, however, Germany has done what Britain did in 1962. The change will encourage increased owner-occupancy (which is relatively low), but future pro-spective tenants may suffer. If the 'pure' owner-occupied house becomes the norm, and tenanted accommodation becomes scarce, West Germany may begin to acquire the 'housing problem' of Britain or the USA. At the moment, this is only a small cloud on the horizon. It is nevertheless disquieting that

German politicians are providing Anglo-American-style tax incentives for 'one's own four walls' without any apparent appreciation of the defects - as well as the virtues - of a 'home-owning democracy'.

'Betterment' Taxes

This category of taxes has a history almost as long as the property tax; a sea wall on Romney Marsh was financed by levies on landowners in 1250. 'Betterment' originally meant the rise in land values resulting from a specific public investment. An assessment was made of this 'betterment' as the basis for a local tax, used to finance the investment. This type of tax ran into difficulties in late 19th-century England when it was applied to the financing of road-widening in cities. The 'betterment' was spread over a wide area, and mixed with 'worsement' for the properties immediately next to a widened road (Turvey, 1957). In the USA, taxes of this sort, known as 'special assessments' also have a long history, but (until recently) had also fallen into disuse. The Dutch version - as Barrie Needham points out - is also rarely used.

When clear-cut 'betterment' results from specific public investment - e.g. a flood-control system - there is still a case for a local 'betterment' tax. Two American writers have, indeed, suggested that 'special assessments' could play a major role in financing subways (underground railways) and similar infrastructure (Hagman and Misczynski, 1978, Chap. 12). There would, however, be problems. It is difficult to decide how much adjacent property has risen in value because a new subway line has been built, and homeowners would complain if they had to pay more taxes merely because of a railway which they might not even use. The Ausgleichs-betraege levied in the German urban renewal areas can also be regarded as an example of this type of tax - although hardly a brilliantly successful one. In general, the application of 'betterment' taxes, in the strict sense, seems likely to remain very limited.

Infrastructure Charges, Exactions etc.

When new residential districts are built, public services have to be provided. The cost can either be borne out of the general revenue of the local community (including the property tax on the new houses) or loaded onto the new development. In the latter case, there is a wide spectrum of possible arrangements. There are 'infrastructure charges' related to the cost of public infrastructure (TLE, Aufschliess-ungsbeitraege, 'special assessments'). A second method is to require the developer to donate land to the community, or build community facilities (participations in France, 'planning gain' in Britain, 'exactions' in the USA). A third approach, mainly found in the USA - the 'impact tax' - is a straight-forward local tax on new construction, e.g. the 'bedroom tax', on each bedroom built.

The case for imposing a charge on new development is partly one of equity; that the costs of associated public services should be borne by the sellers of land and the buyers of the new property, rather than by the community as a whole. It is also one of practicality; if no charge is made, the infrastructure may not be provided by financially hard-pressed public utilities, and development will be hampered. But if - as Dowall suggests is happening in the USA - charges on new developments are used as a substitute for general taxation, there is a danger of restricting the supply of housing, and forcing up its price.

There has also been controversy on whether local authorities should impose clear-cut, legally-defined charges, or be allowed to engage in 'horse trading'. Germany imposes a formal charge; Britain, the USA and, in earlier years, France have relied on 'horse trading', which in the ultimate can become the sale of development permits. The auctioneering of permits has even been advocated by one economist (Pennance, 1967) but, as a general rule, this seems unrealistic and undesirable. The experience in France and Britain of the more extreme forms of 'planning gain' suggests that it is not an ideal system. In France, at least, there has been a move to a more formalised arrangement. Although there will always be a role for haggling, it should probably be within legally-defined limits.

Taxes on Development Value

The three taxes imposed by Labour Governments in Britain were described as 'recouping betterment', and one of them was called a 'betterment levy'. The use of the term was, however, misleading. The taxes were not levied on owners who had benefited from specific public investment. They were levied only when there was development (or a sale of property with implications for development), and on any type of development. To describe such taxes as taxes on 'betterment in a wider sense' is misleading.

This is not to say that a tax on development value is necessarily unjustified. Since it is a form of income, there is as good a case for taxing it as any other form of income - and as good a case for avoiding prohibitive taxation. Britain's forty years of experience confirms what a moment's thought might have indicated - that taxes of 80-100 per cent will throttle the land market, but that a tax of around 30 per cent (with de minimis provisions) can be quite tolerable. Taxes on development value, however, are - contrary to popular belief - not a money-spinner; the receipts have always been minute compared with those from the property tax, let alone from Income Tax, VAT etc. If such a tax is justified, it is because the sight of a few individuals making large untaxed gains is widely regarded as unfair, and can discredit the entire 'free enterprise' system.

DEVELOPMENT CONTROL AND PLANNING

Two controversial issues in several of the countries studied are;

(a) whether sufficient land has been allocated for housing, and
(b) the status and role of development agencies in urban renewal.

On the former question, planners and developers frequently take different views, and it is difficult to reach a verdict. In two countries, however, (Yugoslavia and the USA) one can assert with some confidence that at times insufficient land was allocated. In Yugoslavia, the existence of 'wild suburbs' indicates that the formal sector was unable to supply sufficient housing for the rapidly expanding demand. Georgia Butina argues that, in such a case, the authorities should accept self-building, and help people to build their own houses, rather than pretending that this sector does not exist. In the USA in the 1970s, the restriction on development, in the name of environmental conservation, caused a sharp rise in house prices - in the growing metropolises - and was harmful to people on low incomes. In France, Germany and Britain the evidence is less clear. The level of land prices in Germany and France suggests that a reasonable balance has been struck. In Britain, there may be a case for allocating more land for housing in parts of Southern England, but not elsewhere.

The authors of the national chapters make it clear, however, that the 'housing problem' faced by a section of their population is not primarily a problem of aggregate land availability. It is a problem of poverty; of (especially in Britain) a distorted mix of tenure, resulting from fiscal and legal discrimination against tenancy; of (especially in Britain and the USA) the social and physical breakdown of older residential areas. These problems are exacerbated when new housing is, in effect, reserved for better-off people. The clearest case used to be 'exclusionary zoning' in US suburbs, where minimum lot requirements were used to prevent the construction of cheap housing. Europe now seems to be emulating the USA, although less blatantly. The tendency to 'social polarisation' is particularly marked in Britain, which has had a sharp cut-back in subsidised housing - which was in any case geographically segregated - and an exceptionally

severe 'urban breakdown'. But a similar tendency has been
noted in France, and in the more monolithic of the Dutch
housing estates (e.g. Bijlmermeer, Amsterdam). On the
whole, however, a more active land policy and a healthy
'social housing' sector have countered the tendency in the
Netherlands and Germany.

Urban Renewal

The most intractable aspect of development planning has
been, and still is, the renewal of old residential neighbourhoods. It is intractable because it combines the economic
problem of the changing locational needs of business; the
physical problems of the obsolete housing, road layouts and
industrial plant; and the social problem, pinpointed sixty
years ago, that;

> 'As families and individuals prosper, they escape from
> this area into Zone 2 and beyond, leaving behind as
> marooned a residium of the defeated, leaderless and
> helpless' (Burgess, 1925, quoted in Hallett, 1979).

The renewal of such areas often requires intervention in
the land market by public or quasi-public authorities, because
of the complex and obsolete pattern of landholding - which
may be resistant to change. At the present time, the international trend is for 'partnerships' between the public and
the private sector. Several international conferences have
been held on the theme, and there is a tendency to think
that the organisations set up in different countries - often
with similar names - have the same function. This, however,
is not the case. The US groups are primarily associations of
local businessmen and representatives of interest groups,
operating with the support of the city government, but with
no statutory powers. The German Sanierungstraeger (if they
are not the local government itself) are non-profit-organ-
isations, acting as agents for the local government. The
French SEM are somewhat similar. The Urban Development
Corporations which are being set up by the British Government are quite different. They consist of nominees of the
central Government, and they replace local government. They
both carry out land purchase and subsequent development
and give themselves 'planning permission', thus constituting
not so much judge and jury as plaintiff and judge. They have
been created because the British Government, alone among
the countries studied, regards the governments of the big
cities as the cause of inner-city decay, and is determined to
eliminate their influence. The UDCs bear no responsibility to
the local electorate, and tend to be judged by the central
Government by the amount of development they achieve; there
is thus a built-in tendency to ignore social problems.

As a background to public action, well or badly conceived, the legal and tax system should not impose barriers to renewal, as it sometimes does. In Britain, for example, 'existing use' rights exist in perpetuity, even if the land has been vacant for decades. Such rights should lapse after a specified period of vacancy, thus reducing the 'market value' basis for compensation (Chisholm and Kivell, 1987).

The Urban Revolution and Local Government

In all the countries studied, the 1980s have seen new problems of urban planning arising from the dispersal of economic activity and urban settlement, which Dowall rightly describes as a 'revolution'. This change raises problems for the organisation of local government. Detailed planning is best carried out in small areas, but some strategic problems may need planning over large areas. Should local government be organised in small units, which are close to local problems, or in large units, which may be remote and top-heavy? Or is it possible to get the best of both worlds through a multi-tiered system of government?

The USA has largely retained earlier boundaries, so that many suburbs of large cities have separate governments from 'the city'. Yugoslavia has a remarkably similar system. France and Germany usually have single governments for large conurbations, but France, in particular, has large numbers of very small rural communes (35,000 communes, as against 8,500 in West Germany, which has half the land area). France and Germany have attempted to resolve the problem of co-ordination through regional government, with a fair measure of success. The creation of large local government areas does not necessarily solve the problem. Britain (outside London) has the largest local authorities in Western Europe - and the most powerless. It has no regional governments, and it has created, and abolished, metropolitan governments for the major conurbations in little more than a decade. It is hardly surprising that there are complaints about a lack of strategic planning - even from developers. This international experience provides no patent remedies, but it suggests that;

(a) the financing and powers of local authorities are more important than their size,

(b) great caution is advised in altering local government boundaries, and 'organic' change arising out of agreements between neighbouring authorities is often preferable to solutions imposed from above.

CONCLUSIONS

In conclusion, I will venture to draw a dozen lessons from the experience of the countries studied.

1.   Some measure of public participation in the land market (through compulsory purchase, pre-emption, land-banking etc.) can help to achieve some social and town planning objectives; a monopoly, however, is often unnecessary, and is likely to be undesirable in the type of small-scale development and re-development which will predominate over the next two decades. Public landbanking has in the past been a good financial investment for local authorities, but it is not a 'get-rich-quick' device, especially under present conditions.

2.   Planners have made mistakes, and need to approach their work with considerable humility; on the other hand, laisser-faire in urban development does not produce satisfactory results.

3.   To be effective and satisfactory in the long run, any system of town planning and public land management must be locally based, politically acceptable, and able to operate with continuity. Independent agencies can often serve a useful purpose, but only if they must operate as agents of representative local government.

4.   Both a 'free-for-all' and a uniformly discouraging policy towards housing (and commercial) development produce un-satisfactory results.

'Constraint and encouragement are the two essential ingredients of a regional policy. A policy of constraint in one area can only be successful if accompanied by a policy of encouragement in viable alternative locations' (Dijkstra, 1986).

5.   Any development control system has to strike a balance between general, impersonal rules and discretionary power. A completely 'rule of law' system is hardly feasible, while a completely discretionary system is open to abuse, and will eventually be abused.

6.   Land and housing policy should be designed to encour-age a 'fine grained' urban texture, with a variety of different types and tenures (and, if possible, ages) of houses. Such a pattern is less likely to suffer from physical and social decay than one in which a single public body develops - and, worse still, manages - a large housing area.

7.   Some form of subsidised rented housing is necessary if everyone is to be ensured a reasonable level of housing. It does not have to be 'council housing', but existing institutions should not be destroyed before others are able to cope.

8.   No particular technique or institution concerning land tenure (site value taxation, the leasehold system, pre-emption etc.) should be regarded as a panacea. Similar outcomes can be achieved by a variety of techniques and institutions, depending on the way in which they are operated.

9.   Legislative processes need to be slow, as a result of 'checks and balances'. A system which allows 'instant legislation' does not produce 'strong government', but rather a disruptive cycle of badly drafted legislation.

10.   In framing land taxes, the aim should not to be to find the 'ideal' solution, but to strike the best balance between different, and sometimes conflicting, objectives of equity and efficiency. A market system is fuelled by profit. If, therefore, one wishes to preserve a market system in land transactions, any taxes on land sales or development should not be at rates above 30-40 per cent of the gains actually made. It is, however, perfectly feasible to impose taxes at below this rate. Conversely, some small fiscal 'sweeteners' can greatly ease the prohibition of undesired types of development, while a combination of 'stick and carrot' is needed for vacant inner-city land.

11.   There is a strong case, on grounds of equity and tax neutrality, for taxing the 'imputed rent' of homeownership. If, however, a tax on imputed rent is abandoned, then any general subsidies or tax concessions for homeownership should also be abandoned. There are, however, powerful political pressures from homeowners to subsidise homeownership, in a way which is no longer justified.

12.   There is a strong case, on grounds of equity and resource allocation, for maintaining a local property tax (on housing, business premises and vacant land in urban areas), but there are political pressures for reducing or eliminating it. In the absence of a scheme for sharing national taxes among local authorities, the outcome is liable to be either a restriction in the supply of necessary public services, or a shift in the incidence of taxation so as to bear more heavily on the poor.

Finally, 'The issues involved here are of great complexity, and no perfect solution is to be expected' (Hayek, 1960).

CONCLUSION

REFERENCES

Brittan, S. (1973) Capitalism and the Permissive Society, London, p. 130

Chisholm, M. and Kivell, P. (1987) Inner City Waste Land, Hobart Paper 108, Institute of Economic Affairs, London

Delafons, J. (1969) Land-Use Controls in the United States, MIT: Cambridge, Mass.

Dijkstra, P. (1986) 'Commercial Property Development in Areas of Planning Restraint', in R.J. Towse (ed.) Industrial/Office Development in Areas of Planning Restraint, Kingston Polytechnic, London

Edgeworth, F.Y. (1925) Papers Relating to Political Economy, Vol. 2, p. 197

Frieden, B. (1979) The Environmental Protection Hustle, MIT, Cambridge, Mass.

Goldberg, M.A. (1977) 'Housing and Land Prices in Canada and the US', in L.B. Smith (ed.), Public Property: the Habitat Debate Continued, The Fraser Institute, Vancouver

Hagman, D. and Misczynski, D. (1978) Windfalls for Wipeouts: Land Value Capture and Compensation, Chicago

Hallett, G. (1977) Housing and Land Policies in West Germany and Britain, London

Hallett, G. (1979) Urban Land Economics: Principles and Policy, London

Hayek, F.A. (1960) The Constitution of Liberty, p. 350

Jacobs, J. (1962) The Death and Life of Great American Cities, London

Lefcoe, G. (1979) Land Development in Crowded Places; Lessons from Abroad, The Conservation Foundation, Washington, DC

Lefcoe, G. (1981) 'California's Land Planning Requirements: The Case for Deregulation', in A Conference on Land Policy and Housing Development, Monograph 81.5, Lincoln Institute, Mass. p. 264. Also Southern California Law Review, 1981

Logan, T. (1976) 'The Americanisation of German Zoning', Journal of the American Institute of Planners

Masser, I. (1986) 'The Transfer of Planning Experience between Countries', in I. Masser and R.H. Williams (eds.), Learning from Other Countries, Norwich

Meade, J.E. (1978) The Structure and Reform of Direct Taxation, Institute of Fiscal Studies, London

Norton, A. (1983) 'The Province of Ontario, Canada: the Municipality of Metropolitan Toronto', in The Government and Administration of Metropolitan Areas in Western Democracies, University of Birmingham

Pennance, F.G. (1967) Housing, Town Planning and the Land Commission, Hobart Paper 40, Institute of Economic Affairs, London

Prest, A.R. (1981) The Taxation of Urban Land, Manchester

Roberts, N.A. (1977) 'Canada; Small-Scale Government Land Development and Large-Scale Private Developers', in The Government Land Developers, N.A. Roberts (ed.), Lexington, Mass.

Sandford, C.T. (1975) An Annual Wealth Tax, London

Siegen, B.H. (1972) Land Use Without Zoning, Lexington, Mass.

Stanley, P. (1987) All Change, South East Development Strategy, Stevenage, UK

Strong, A.L. (1979) Land Banking; European Reality, American Prospect, Baltimore

Topping, K.C. (1981) 'San Bernadino County's Planning Process Reforms', in A Conference on Land Policy and Housing Development, Monograph 81.5, Lincoln Institute, Cambridge, Mass., p. 181

Turvey, R. (1957) The Economics of Real Property, Chap. IX, London

For Product Safety Concerns and Information please contact our EU
representative GPSR@taylorandfrancis.com
Taylor & Francis Verlag GmbH, Kaufingerstraße 24, 80331 München, Germany

9 780367 772062